Case Studies in Environmental Science

Robert M. Schoch
Boston University

West Publishing Company
Minneapolis/St. Paul New York Los Angeles San Francisco

Cover Image: Wood Ranch, CA.
©1996 Dean DeChambeau

WEST'S COMMITMENT TO THE ENVIRONMENT

In 1906, West Publishing Company began recycling materials left over from the production of books. This began a tradition of efficient and responsible use of resources. Today, up to 95% of our legal books and 70% of our college texts and school texts are printed on recycled, acid-free stock. West also recycles nearly 22 million pounds of scrap paper annually—the equivalent of 181,717 trees. Since the 1960s, West has devised ways to capture and recycle waste inks, solvents, oils, and vapors created in the printing process. We also recycle plastics of all kinds, wood, glass, corrugated cardboard, and batteries, and have eliminated the use of Styrofoam book packaging. We at West are proud of the longevity and the scope of our commitment to the environment.

Production, Prepress, Printing and Binding by West Publishing Company.

Printed in the United States of America
03 02 01 00 99 98 97 96 8 7 6 5 4 3 2 1 0

ISBN 0–314–20397–4

CONTENTS

Preface.....v

SECTION 1. Environmental Principles

1. Early Admonitions from George Perkins Marsh.....3
2. The Mystery of Easter Island.....5
3. Does Environmental Reporting in the Popular Media Tend to be Biased and Sensationalized?.....9
4. Beach Nourishment: A Worthy Endeavor or a Waste of Money?.....13
5. Diversity and Ecosystem Productivity.....15
6. Will a Technological Fix Work, or Do We Need Fundamental Social Change?.....17
7. The Costs of Saving Lives.....21
8. Differing Approaches to the Population Problem.....23
9. Famine and Overpopulation in Ethiopia.....27

SECTION 2. Problems of Resource Depletion

10. Making the Decision: Building a Highway to a Hospital or Protecting an Endangered Beetle.....33
11. The Wildlands Project.....37
12. The Environment Held Hostage: The Gulf War Experience.....39
13. The *Exxon Valdez* Oil Spill.....43
14. The Dangers of Conservation.....47
15. Circumventing Nuclear Waste Disposal By Reclassification.....49
16. So Just How Dangerous is Plutonium?.....51
17. How Easy is It to Build a Nuclear Bomb?.....53
18. A Uranium Plant Disguised.....57
19. Future Water Wars.....59
20. Unwanted Species in Hawaii.....61
21. Killing Elephants in an Overcrowded Park.....65
22. Should Gulls be Poisoned to Make Way for Other Birds?.....69
23. Should Mountain Lion Hunting be Legalized in California?.....71
24. Overcrowding in the National Parks--Yosemite National Park as an Example.....73
25. An Excess of Deer.....77
26. Saving the Galapagos.....81

27. Smallpox: Should It Be Eradicated Completely? 85
28. Saving Wood 89
29. Spotted Owls and Old-Growth Forests in the Pacific Northwest 93

SECTION 3. Problems of Environmental Degradation

30. The Rise and Fall of DDT 99
31. Computer Chips, Carcinogens, and Health Risks 103
32. Are All Risks Equivalent? 107
33. Dealing with Lead Paint 111
34. The Subtle Dangers of Synthetic Chemicals 115
35. The Controversy over Water Fluoridation 119
36. The Dangers of Wood Smoke 123
37. The Potential Dangers of Electromagnetic Fields 125
38. Should Nuclear Bomb Testing Be Resumed? 129
39. Methyl Bromide and the Ozone Layer 133
40. Ultraviolet Damage in Canadian Lakes 137
41. The Dump in Wellesley, Massachusetts 141
42. Recycling Disposable Diapers 143
43. The Wastefulness of Reuse and Recycling 145
44. Should Toxic Waste Attract More Toxic Waste? 147
45. Liability for Pollution from a Superfund Site 151
46. Nuclear Reactors at the Bottom of the Oceans and in Space 153
47. The Garbage Sifters of Cairo 155
48. Truth in Advertising and Campaigning 157

SECTION 4. Social Solutions

49. Green Dollars--A Vancouver Experiment 163
50. Profits or Environmental Responsibility? 167
51. Green Consumerism, or No Consumerism? 169
52. Green Markets in Mexico 173
53. Aztec Cannibalism 175
54. Thoreau at Walden Pond 179
55. The Life and Death of Chico Mendes 183
56. Catastrophic Environmental Predictions from the First Earth Day (1970) 187
57. Tree Spiking 191
58. Legal Rights for Trees and Streams 193
59. The Takings Concept: When is Just Compensation to a Property Owner Required? 197
60. Easterbrook's Concept of a New Nature 201

Preface

This collection of sixty case studies is intended to supplement the "Case Studies," "Issues in Perspective" boxes, "Prologues," and text of *Environmental Science: Systems and Solutions* by Michael L. McKinney and Robert M. Schoch (West Publishing Company, 1996). These case studies can be used by an instructor as enrichment material at any time during an environmental science or environmental studies course. Here they are arranged to follow the general order of material covered in *Environmental Science*. Accordingly, the case studies are grouped into four broad categories corresponding to the four sections of *Environmental Science*: Environmental Principles, Problems of Resource Depletion, Problems of Environmental Degradation, and Social Solutions.

Acknowledgments

I thank Jerry Westby, college editorial manager for West Publishing Company, and Dean DeChambeau, West Publishing Company developmental editor, for their help and encouragement during the writing of these case studies. I also extend my sincere appreciation to my wife, Cynthia, and sons, Nicholas and Edward, for their patience during the writing of this volume. In addition, Cynthia read all of the case studies and offered many helpful comments.

Robert M. Schoch, Attleboro, MA

Section 1

Environmental Principles

-1-

Early Admonitions from George Perkins Marsh

In 1864 a book entitled *Man and Nature; or, Physical Geography as Modified by Human Action* was published by the American diplomat George Perkins Marsh (1801-1882). Marsh was a native Vermonter, but had travelled extensively throughout Europe and the Near East. Studies in these areas, particularly France and the Middle East, were documenting the harmful effects of deforestation. Synthesizing such research with his own observations, Marsh argued that declines of whole empires and civilizations followed destruction of the woodlands upon which they ultimately depended. Deforestation meant not only a lack of wood, but caused the destruction of the soil, reduced and contaminated the water supply, and caused devastation among the native flora and fauna. Marsh challenged the notion of an inexhaustible Earth, or an Earth that could always heal itself in the face of humanity's onslaughts.

Marsh did not argue against humankind's intervention in, and transformation of, nature. But he did suggest it must be done with knowledge and foresight, if irreparable damage and negative consequences were to be avoided. Marsh's work was extremely influential, and helped lay the groundwork for the late nineteenth-century conservation and preservation movements.

The following passage is quoted from *Man and Nature*:

> He [civilized, modern man] has felled the forests whose network of fibrous roots bound the mould to the rocky skeleton of the earth; but had he allowed here and there a belt of woodland to reproduce itself by spontaneous propagation, most of the mischiefs which his reckless destruction of the natural protection of the soil has occasioned would have been averted. He has broken up the mountain reservoirs, the percolation of whose waters through unseen channels supplied the fountains that refreshed his cattle and fertilized his fields; but he has neglected to maintain the cisterns and the canals of irrigation which a wise antiquity had constructed to neutralize the consequences of its own imprudence. While he has torn the thin glebe [that is, soil] which confined the light earth of extensive plains, and has

destroyed the fringe of semi-aquatic plants which skirted the coast and checked the drifting of the sea sand, he has failed to prevent the spreading of dunes by clothing them with tribes of animated nature whose spoil he could convert to his own uses, and he has not protected the birds which prey on the insects most destructive to his own harvests. . . .

We have now felled forest enough everywhere, in many districts far too much. Let us restore this one element of material life to its normal proportions, and devise means for maintaining the permanence of its relations to the fields, the springs and rivulets with which it waters the earth. The establishment of an approximately fixed ratio between the two most broadly characterized distinctions of rural surface--woodland and plough land--would involve a certain persistence of character in all the branches of industry, all the occupations and habits of life, which depend upon or are immediately connected with either, without implying a rigidity that should exclude flexibility of accommodation to the many changes of external circumstances which human wisdom can neither prevent nor foresee, and would thus help us to become, more emphatically, a well-ordered and stable commonwealth, and, not less conspicuously, a people of progress.

Questions

1. The passage quoted above was written over a hundred and thirty years ago. Are we still facing the same types of problems? Have we taken the preventive measures that Marsh suggested?

2. In the first paragraph of the Marsh quote given above, who or what is the "wise antiquity" that he mentions?

3. How does Marsh anticipate the modern sustainability movement?

4. If Marsh were alive today, how do you think he would view our current environmental situation?

Sources

Bowler, Peter J., 1992, *The Norton History of the Environmental Sciences*. New York: W. W. Norton and Company.

Marsh, George Perkins, 1864, *Man and Nature; or, Physical Geography as Modified by Human Action*. New York: Charles Scribner.

Nash, Roderick Frazier, 1990, *American Environmentalism: Readings in Conservation History (third edition)*. New York: McGraw-Hill Book Company.

-2-

The Mystery of Easter Island

Easter Island (Isla de Pascua), located over 2000 miles off the coast of Chile in the southern Pacific ocean, is a triangular shaped island measuring only fifteen miles along its base and less than eight miles across at its widest point, composed of the remains of several extinct volcanoes. Due to its extreme distance from any other land, the indigenous biota of Easter Island was quite sparse, consisting primarily of a few dozen types of plants, various insects, and small lizards.

When a Dutch admiral and his crew discovered Easter Island on Easter Sunday, 1722, he found an island inhabited by approximately 3,000 native Polynesians. These natives lived in squalid poverty, eking out an existence by cultivating sweet potatoes in the poor soil and raising chickens. They lived in reed huts or caves, had virtually no wood (for there were almost no trees on the island), had no boats that were ocean-going and thus could not leave their island, and continuously engaged in warfare with one another. Given their poverty and lack of food supplies, the Easter Islanders even took part in cannibalism. The Dutch were not generally impressed and no Europeans visited the island again for almost 50 years (Spanish sailors stopped by in 1770). Throughout the eighteenth and nineteenth centuries the human population declined on Easter Island, due both to internal causes and the fact that periodically European or American ships would come to the island and take away slaves. The final blow came in 1877 when Peruvians arrived and carted away virtually all that remained of the population, leaving behind only about 110 children and elders. Later the island was taken over by Chile and used primarily as a giant sheep ranch.

There was more to Easter Island than just poverty-stricken, warring, cannibalistic clans. Scattered over the island are massive stone statues that weigh tens of tons each and when erect stand twenty feet high on average (most have been toppled). There were more than 600 such statues on the small island! Associated with the largest of the statues were massive stone platforms. This seemed to be evidence of an early, advanced civilization that had inhabited, or at least visited, the island. Surely the natives could not have constructed such monuments. The "mystery" of Easter Island was born.

Early archaeologists and anthropologists had no good explanation as to how, why, and by whom the statues had been built. They did know they were carved from a quarry in the volcanic rock on the island, for some remained in place only partly carved out. From the quarry they were somehow transported and erected at their present locations. Such feats were clearly beyond the capabilities of the natives discovered by the Dutch in 1722. In the twentieth century theories about the Easter Island mystery proliferated, ranging from notions that ancient monumental stone builders had visited the island from South America, to more fanciful ideas focussed on lost Pacific civilizations that had sunk beneath the ocean (a Pacific version of the Atlantis myth) or tales of visits from extraterrestrials. Slowly, the true story behind Easter Island was pieced together, and although it may be more mundane than some of the wilder speculations, it is also more instructive.

The first human settlers on Easter Island were Polynesians who arrived in the fifth century A.D. The original settlers, travelling on ocean-going double canoes, numbered perhaps two dozen and carried with them chickens and sweet potatoes. On Easter Island sweet potatoes could be cultivated, and chickens raised, with very little effort. Although the settler's diet may have been a bit dull, it was nutritionally balanced and left them with plenty of free time. There was also an abundance of trees with which to build houses and canoes, and make fishing nets, cloth (actually a type of paper by modern standards), and other important items.

Over the next thousand years the Polynesian population steadily increased, until it reached an estimated peak of 7,000 in about 1550. The Easter Islanders developed an elaborate social organization based on extended families and clans. Competition for prestige and power between the clans, combined with an abundance of free time, seems to have led to an elaborate civilization which culminated in the erecting of the huge stone statues and accompanying stone platforms.

Carving the statues was not a difficult feat technologically--as long as one had plenty of time it was easily accomplished using primitive stone tools. The real feat was to move the statues from the quarry to the farthest corners of the island. The islanders had no horses, oxen, or other large animals to help with the burden, so the statues had to be moved by human ingenuity and muscle alone. Their solution was to use tree trunks as rollers. They apparently made virtual roads of tree trunks and passed the statues along them, almost like a track or conveyer belt in a modern factory. Over the centuries the islanders began cutting trees for this purpose faster than they could naturally grow back; the end result was massive deforestation of the island. By 1500 the situation was extremely critical, and a century later the island was virtually treeless. Without trees the statues could not be moved; the end came so fast that some statues were literally abandoned in the quarry.

Deforestation did not just impact the raising of statues and the important religious, social, and ceremonial functions that surrounded the statues. It also cut at the core of the essential economic and agricultural base of the island. Without trees the people could not build timber houses, carve sea-worthy canoes, or produce their fishnets and cloth. Furthermore, as a result of the massive deforestation the soil of the island eroded and became impoverished of nutrients. Crop production declined, food and all other resources became increasingly scarce, tensions rose between clans, and warfare become prevalent on the island. War only worsened the situation, sapping the people's strength and resources. The once magnificent Easter Island civilization, arguably one of the highest achievements of

Polynesian culture, totally collapsed. It was in this condition that the Easter Islands were discovered by eighteenth-century European navigators.

Questions

1. The rise and fall of Easter Island civilization has been viewed by some scholars as representative of the cycle of many civilizations. What other civilizations have followed a similar path? Was their collapse due to environmental degradation analogous to that of Easter Island?

2. It has been suggested that by 1550 the Easter Islanders must have clearly realized the ecological destruction, particularly the deforestation, that they were causing--yet they persisted in their activities until the very end. How do you explain their actions? Why didn't they change their behavior before it was too late?

3. Do you think it is fair to compare Easter Island to the planet Earth as a whole? Both are small "islands" in vast expanses of "ocean" that bear only limited supplies of resources and are inhabited by humans who cannot leave their respective homes. If we are not careful, is it possible that the story of Easter Island will be the story of Earth? What lessens should humanity be learning from the sad tale of the Easter Islanders?

Source

Ponting, Clive, 1992, *A Green History of the World: The Environment and the Collapse of Great Civilizations*. New York: St. Martin's Press.

-3-

Does Environmental Reporting in the Popular Media Tend to be Biased and Sensationalized?

Many critics of the popular press, such as commercial television shows, daily newspapers, and weekly magazines, charge that reporting on environmental issues is often biased and sensationalized. Such critics are not only found among the "anti-environmentalist" camp, but also include a number of people sympathetic to the environmental movement. One of the more outspoken critics of the popular press when it comes to reporting on environmental issues has been Jane S. Shaw. Shaw has an inside view of the press--she was an associate economics editor of *Business Week* from 1981-1984.

Shaw came down especially hard on *Time* magazine when, in 1989, the editors of *Time* temporarily abandoned their concept of "Man of the Year" for an issue devoted to the "Planet of the Year." The planet, of course, was Earth and the magazine carried a number of articles describing then current environmental issues. The *Time* writers and editors not only attempted to describe the problems the planet faced, but suggested specific recommendations that might be implemented in order to solve the problems. For instance, after discussing the deterioration of the ozone layer by chlorofluorocarbons, *Time* called for a complete ban on the manufacture of all CFCs. Later in 1989 Charles Alexander, the science editor of *Time*, publicly stated that "as the science editor at *Time* I would freely admit that on this issue [environmental issues as covered in *Time*] we have crossed the boundary from news reporting to advocacy."

Critics such as Shaw charge that too often the popular press reduces complex environmental issues into overly sensationalized, even lurid accounts. All sides of what is often an extremely controversial issue are not reported; fairness and objectivity are lost. A case in point, according to Shaw, would be "global warming"--an issue concerning which many equally competent scientists still disagree. Is global warming really occurring? If so, how much global warming can we expect? What will be the potential consequences? What actions should be taken, if any? Despite scientific uncertainties, the popular press still may take it upon itself to recommend "solutions" after first summarily reporting, in a greatly exaggerated and over-simplified manner, the "problems."

Another case in point, cited by Shaw, is the 1989 media scare over Alar (daminozide) on apples. Alar is a synthetic substance that can be applied to apples in order to regulate their growth, thus keeping them from falling off the tree too early or ripening too fast once picked. The Environmental Protection Agency found that Alar (or more specifically its breakdown product, unsymmetrical 1,1-dimethylhydrazine) ingested in high doses could cause tumors in laboratory test animals, and when this information got out the popular press picked up on the story and suddenly many people were afraid to eat apples or apple products. The Natural Resources Defense Council published a report concluding that thousands of children who eat apples and apple products (such as apple juice and apple sauce) could be at risk of cancer due to Alar contamination. CBS's popular show "60 Minutes" picked up the NRDC's report and widely popularized the conclusions. Complete bans on Alar were called for. All apple growers suffered badly as the public refused to eat any apple products, despite the fact that the federal government estimated that at most 15% of apple growers ever used Alar, and perhaps only 5% used it in 1988. Some family farms reportedly went bankrupt. In response, apple growers announced a voluntary ban on Alar, the manufacturer withdrew it from the American market, and a few years later the EPA banned Alar.

Critics of the Alar scare and its heavy promotion by the media argue that Alar and its by-products are not nearly as dangerous as they were assumed to be; indeed, the World Health Organization does not even classify Alar as a carcinogen. Furthermore, it was contended, the use of Alar in some cases reduced the need for other, potentially more dangerous pesticides. Finally, when the Alar controversy was at its height the consumption of apple products dropped dramatically--not only was this bad for apple growers, but also bad for the school children and others who substituted less healthy snacks for apple products.

Still, even if the Alar threat was overblown by the media, supporters of the ban contend that society is better off without Alar. Alar was used only by a minority of apple growers, and the apple business has done fine without Alar. Furthermore, due to the public outrage over Alar, the EPA became more responsive to the public. Additionally, attempting to avoid bad publicity, pesticide manufacturers began to cooperate with the EPA more closely. All in all, the Alar controversy had many positive, if unintended, outcomes.

But even if we grant that the Alar scare was productive in the end, critics believe it does not justify the often apparently one-sided coverage of this and other environmental issues by the popular press. Such critics believe that the press has a duty to be completely fair and objective, reporting just the "facts" (which often means giving equal weight to both sides of a controversial environmental issue) in a straightforward manner without bias or over-simplification, and without making suggestions as to how the problem should be solved.

Shaw believes that the popular media strays from the ideal outlined above because for most people news is essentially a form of entertainment, so ultimately the most "entertaining" news will be reported (or even fabricated, through the oversimplification of complex issues). Reporters, producers, and editors will generally deliver the product that their public demands, for they must sell copies of their magazine or garner strong television ratings. People are not generally interested in long, drawn-out, complicated stories; they are not worried about having an in-depth understanding of the issues. Rather, they desire a superficial, entertaining approach. A good story has to have a clear, simple story line. The best stories are straightforward and can be told in terms of good versus evil (for example, the good, altruistic environmentalists fighting the big, bad, greedy pesticide-producers). Ambiguity does not appeal to most people, and thus does not attract readers or viewers. When stories are painted in terms of good versus evil, it is an easy thing for the reporter or editor to weigh in on the side of good.

Questions

1. Do you think that reporters and editors can, or even should, be "totally objective?" Why or why not? Does it perhaps depend on the subject matter involved in a particular story?

2. What is wrong with reporters or editors suggesting solutions to problems that they perceive? Is this any different than newspapers writing editorials or endorsing candidates for elected office?

3. What is your view of the Alar controversy? Do you think the media handled themselves responsibly, or were they just looking for readership and ratings and therefore got on the Alar scare bandwagon?

4. Can you cite some recent examples of biased media reporting on environmental issues? How about unbiased, objective media reporting?

5. Do you believe that the public is ultimately to blame for the way the popular media handle complex environmental issues?

Sources

Easterbrook, Gregg, 1995, *A Moment on the Earth: The Coming Age of Environmental Optimism*. New York: Viking. [Comments on Alar.]

Frank, Irene, and David Brownstone, 1992, *The Green Encyclopedia*. New York: Prentice Hall General Reference. [Comments on Alar.]

Harte, John, Cheryl Holdren, Richard Schneider, and Christine Shirley, 1991, *Toxics A to Z: A Guide to Everyday Pollution Hazards*. Berkeley: University of California Press. [Comments on Alar.]

Shaw, Jane S., 1992, "Is Environmental Press Coverage Biased?" In *Rational Readings on Environmental Concerns* (edited by J. H. Lehr), pp. 474-484. New York: Van Nostrand Reinhold.

Time (January 2, 1989).

-4-

Beach Nourishment: A Worthy Endeavor or a Waste of Money?

Beach erosion is a serious problem, both environmentally and perhaps even more so economically, in many coastal areas, particularly along the Atlantic and Gulf coasts of the North America. Many different methods have been attempted to deal with such problems, such as building jetties, breakwaters, seawalls, and artificial reefs. These methods seek primarily to protect and preserve a beach from further erosion; another approach to the problem is to try to restore and build up the beach with new sand, a procedure commonly known as beach nourishment.

In the typical beach nourishment project a large deposit of sand is located offshore, dredged, and loaded onto barges. The sand is brought to the beach area that is to be restored, mixed with water, and the sand-water slurry is pumped onto the beach. The sand grains settle out and are added to the sand already on the beach. A single beach nourishment project can cost tens of millions of dollars. Between 1950 and 1993 the Army Corps of Engineers oversaw 56 major beach nourishment projects financed in part by the Federal Government. The total cost for these projects was nearly $1.5 billion, with the government picking up almost $900 million of the tab.

The first beach nourishment project in the United States was carried out in 1922 when New York City decided to enlarge its Coney Island beach. Arguments over the efficacy of beach nourishment were raised then, and the debate has been going on ever since. Opponents of beach nourishment suggest that such efforts are often futile; in most cases much of the new sand is quickly eroded away, especially if a large storm hits the beach. At best, such beach nourishment projects serve primarily to protect (even if only temporarily) the investments of coastal property owners--and therefore when heavily funded by the federal government amount to a subsidy to such owners at the general taxpayers' expense. In some cases the projects are a total waste of money. Citing specific examples, critics note that a beach nourishment project in Ocean City, Maryland which cost nearly $70 million was nearly destroyed by severe storms during the fall and winter of 1991-92, just after the project was completed. In the case of nourishment on Folly Beach, South Carolina, 2.4 million cubic yards of sand were placed on the beach in 1993 so as to widen it by close to 60 yards; however, a year later about 20 yards had already been lost.

Proponents suggest that many of the beach nourishment failures cited by the critics are actually successes. It cannot be expected, they argue, that all of the new sand will remain in place. Rather, the new beach has to be shaped by natural forces and reach a relatively stable equilibrium. Or, in some cases, a beach must be continually renourished every few years. Referring to Folly Beach, 40 extra yards of width a year later is still a significant addition to the beach. Relative to the Ocean City project, if the beach nourishment had not been done before the unexpected storm then the damage would have been even worse. As it was, because of the beach nourishment project many hotels and other buildings were saved as the storm's energy worked on the sand that had been newly laid on the beach rather than attacking the natural coastline. The bottom line, according to proponents, is that the money spent on beach nourishment projects has saved many times the value of expensive coastal real estate.

Questions

1. What do you think of beach nourishment projects? From a long-term, geological perspective beaches are extremely shortlived, ephemeral phenomena. Should humans be artificially tampering with the coastline, or should the beaches be allowed to develop naturally?

2. Do you think the money spent on beach nourishment projects is justified? Explain your answer.

3. Given that a certain beach nourishment is to be carried out, who should pay the expenses? Should this be the responsibility of the owners of coastal properties? Should the federal government help finance such projects? Why or why not?

Source

Dean, Cornelia, 1996, "Is It Worth It to Rebuild a Beach? Panel's Answer is a Tentative Yes." *The New York Times* (April 2, 1996), p. C4.

-5-

Diversity and Ecosystem Productivity

As far back as 1859 in *On the Origin of Species* Charles Darwin stated that the more diverse (in terms of number of different species) an ecosystem is, the more productive it will be (in terms of biomass that can be produced and sustained), the more efficiently it will use and recycle resources (such as limiting nutrients), and the more resilient it will be when struck by a disturbance (such as a drought that affects a grassland or forest). Many conservation biologists and ecologists have long assumed that diverse ecosystems tend to exhibit these types of features, but demonstrating this definitively is no easy matter. Limited studies in small, artificial, greenhouse settings seemed to confirm the hypothesis that increased biodiversity correlates with increased productivity, but a major question is whether this is always true. Artificial ecosystems may not precisely mimic those in nature.

One researcher who has been intensively studying this issue is Dr. David Tilman, an ecologist at the University of Minnesota. In studies of the Minnesota prairie Tilman and his colleagues have shown that what Darwin and others have believed all along is basically true. Studying the prairie ecosystem after the severe Midwest droughts of 1987 and 1988, Tilman's team found that areas with higher species diversity returned to their pre-drought productivity much more quickly than areas with fewer species. In a two-year controlled experiment, the researchers first burned (so as to destroy all the vegetation) 147 hundred-square-foot plots of prairie land in the Cedar Natural History Area of Minnesota. They then plowed and planted each plot by hand with from one to twenty-four prairie species native to the area. The number of species, and the particular combination of species, for each plot was chosen randomly. One problem the researchers faced was that seeds from other plants might either still remain in the soil after the initial burning, or be blown into the plot, thus compromising the integrity of the study. Therefore, as the plants grew, any species other than those chosen for the particular plot were weeded out by hand.

After two years the scientists found that the plots with higher biodiversity had indeed produced more plant biomass and had utilized nitrogen, a basic plant nutrient, more efficiently. In contrast, plots with few species produced less biomass and nitrogen tended to leach out of the upper layers of the soil, collecting below the level to which most of the plants' roots reached, thus further limiting plant growth.

In the Minnesota study the positive effects of biodiversity steadily and dramatically increased with the addition of new species up to about ten species. After that, adding still more species did continue to increase productivity, but only slightly for each species added. This suggests to some ecologists that after about ten species the functions of the various species begin to overlap--redundancy enters the ecosystem. The study also did not precisely address why increased biodiversity correlates with increased productivity. Most ecologists suggest that it probably has to do with the fact that each species utilizes resources in a slightly different manner; therefore, collectively, a number of different species can utilize all the resources most effectively. This hypothesis is far from proven, however.

Questions

1. What types of questions have Tilman's prairie studies left unanswered? The 147 plots described above were monitored for only two years--do you think the results might change if they are reevaluated after ten or twenty years? Can research on prairie ecosystems be extrapolated to other types of ecosystems? Based only on the prairie studies should one make generalizations about the effect of biodiversity on ecosystem productivity?

2. The Tilman study focussed on a maximum of twenty-four prairie plant species. Do you think this is the whole story? What about the numerous bacteria and other microbes, fungi, insects, worms, and other organisms found in the soil that are legitimately part of any terrestrial ecosystem?

3. What practical benefits might the Minnesota prairie studies have? Can these results be applied to agriculture or to forest ecosystems? Some crop and greenhouse experiments have produced very high productivities using only a single species at a time--indeed the concept of monoculture was important in the Green Revolution that produced strong increases in world grain yields during the 1960s through 1980s. However, are such monocultures sustainable in the long run? Are natural Minnesota prairie ecosystems sustainable? Should humans strive for both sustainability and high productivity in ecosystems and agricultural systems?

Source

Yoon, Carol Kaesuk, 1996, "Ecosystem's Productivity Rises with Diversity of Its Species." *The New York Times* (March 5, 1996), p. C4.

-6-

Will a Technological Fix Work, or Do We Need Fundamental Social Change?

Among those concerned about environmental issues, there are two major schools of thought concerning how our problems can be solved. One believes that even if many of the assaults to the planet have been brought on by modern technology (for instance, ozone destroying chemicals, artificial radionuclides, and overpopulation due to modern medicine and industrialization), the solutions also lie in modern technology. Future scientific and technical breakthroughs will be the key to solving our environmental problems. The other school of thought holds that we are ill-advised to rely on new, unpredictable, yet to be developed technological fixes; rather, what is needed are fundamental political, economic, and cultural changes among human societies in order to address the current environmental crises.

Those advocating the technofix approach base their analyses in large part on past history. Scientific and technological knowledge has accelerated at an ever increasing pace over the last few hundred years, and especially during the last century. In the 1890s few could have imagined the electronic capability of the 1990s, the fact that humans can make regular trips to space and had already reached the Moon, the use of a metal (uranium) as an energy source, the unprecedented grain yields produced by modern fertilizers, the replacement of woods and metals by plastics for many uses, synthetic clothes and other substances, and so many other developments. Believers in technology place their faith in the idea that before any particular environmental problem becomes unbearable, a scientific and technological breakthrough will occur to mitigate or solve the problem. In this view, this has always been the general rule and there is no reason to doubt it will hold true in the future. Indeed, as scientific knowledge increases at an ever quickening pace (scientific knowledge grows exponentially, just as global human population does), we can predict scientific breakthroughs developing at a faster rate.

Extreme anti-technologists reject the preceding argument. Rather than looking to future technology to solve our environmental problems, they blame technology and industry for those very environmental problems in the first place. If it were not for the modern industrial-technological complex we would not have our current pollution and hazardous waste problems. Synthetic chemical compounds have done more harm than good--now they contaminate the air, soil, and water. Stratospheric ozone is being destroyed by that twentieth-century group of wonder chemicals known as chlorofluorocarbons. The internal combustion engine is largely responsible for the potential of significant global warming. Nuclear power never fulfilled its promise of cheap, virtually unlimited quantities of electricity--yet it has left us with a deadly legacy of radioactive pollution and stockpiled waste. Modern agricultural techniques have significantly increased world food production, but this has perhaps only encouraged the growth of the global human population. Even if certain aspects of science and technology have helped to address particular environmental problems, most of those problems were caused by modern technology in the first place.

Even if we accept that so far science and technology have allowed the world to keep up--to muddle along--and necessary breakthroughs have come as needed; conditions have improved for at least some people, more people are fed and sheltered, and so forth; is this necessarily good? Is this what we really want? As the global population continues to increase more people (in terms of absolute numbers) are living the "good life," but even more people are suffering. Projections for the immediate future do not predict a change in this trend. In 1950 67% of the world's population lived in less developed countries, and 33% lived in the more developed countries (essentially the modern, industrialized nations). By the 1990s these percentages were 77% and 23%, and based on past trends, by 2025 84% of the world's population will live in less developed countries. In 1950 there were only 2.5 billion people inhabiting our planet, there were 5.8 billion in 1996, and by 2025 there will be an estimated 8.5 billion. Of those 8.5 billion, only about 1.35 billion, those in the more developed countries, will receive the major benefits of modern science and technology. The other 7.15 billion will be left behind.

Thus, the anti-technologists argue, based on looking at all aspects of past history objectively, the best we can expect even when we factor projected breakthroughs in science and technology into the equation, is a future where the discrepancy between rich and poor will increase, not decrease. In 2025 the poor will outnumber the rich by a factor of over 5 to 1. By the end of the twenty-first century the poor could outnumber the rich by as many as 10 to 1. The technofixers usually imply that technology and science are making the world better for everyone, but this is just not the case. The trend has clearly been for an ever smaller percentage of the people to share in the fruits of technology even as resources flow from poor to rich nations. And of course even in the rich nations there are many poor citizens who do not reap all the benefits available to their wealthier counterparts.

One can argue that such gross inequalities between the poor and rich of the world cannot continue forever. As the poor increase in number they will resent the rich more and more. Eventually the poor may rebel; international conflict may break out; the tools of mass destruction, such as nuclear weapons (another gift of science and technology), may be used as a method to redistribute the riches and spoils of the world. In the aftermath, how will most people view technology?

Before such a ghastly scenario plays itself out, some anti-technologists argue that we must radically rearrange our social, cultural, and political institutions so as to ensure the more equitable distribution of food, material goods, and services to all peoples of the world. We must not work toward more for the few, but toward sustainability for the whole Earth.

Certain technological and scientific advances may help achieve this goal, but fundamentally it will take a rethinking of our basic values. The solution must come from within humans, not from another miraculous technological breakthrough.

Questions

1. Would you characterize yourself as more of a "technofixer" or an "anti-technologist"? Why? Do you trust technology?

2. In general, do you think modern technology has done more harm or more good? Explain your answer.

3. How have dominant political and social values and attitudes changed over the past century? How have technological breakthroughs affected political and social thinking? For example, once nuclear weapons were developed, could people ever view global international conflicts the same way they viewed them in the pre-nuclear age?

4. Is it easier to develop a new technology or to change people's basic attitudes and ways of thinking? Which is easier for a government in power to direct or control, a particular technology or social unrest and political change?

Sources

Dobson, Andrew, 1991, *The Green Reader: Essays Toward a Sustainable Society.* San Francisco: Mercury House.

Easterbrook, Gregg, 1995, *A Moment on the Earth: The Coming Age of Environmental Optimism*. New York: Viking.

Trainer, Fred, 1985, *Abandon Affluence!* London: Zed Brooks Ltd.

-7-

The Costs of Saving Lives

How much is a life worth? As cruel as this question may seem, in some contexts it is a legitimate question. Funds are limited and one should consider the cost-effectiveness of various safety programs. The cost per life saved of various programs and undertakings has been calculated and catalogued by Cohen (1990) in order to compare the price of increased safety features at nuclear power plants to other ways that money could be spent. According to Cohen's work, the bottom line is that the cost of Nuclear Regulatory Commission safety systems and programs requires the spending of millions to billions of dollars per life saved. In contrast, in America many routine medical tests can save thousands of people from cancer at a cost of less than $100,000 per life saved. An example is a periodic test that screens for certain cancers, such as annual Pap smears for women. If an annual Pap test has one chance in 3,000 of detecting cancer and thus saving a life, and the Pap test costs $30 each, then the cost per life saved is 3,000 times $30, or $90,000 per life.

Improved medical facilities, improved traffic signs, safety devices in automobiles, and so on are all ways that money can be easily spent in America to save lives at no more than a few hundred thousand dollars per life saved. At a global level, many lives can be saved in underdeveloped and Third World nations through immunization programs, disease control, water sanitation, and food distribution--less than $100 per death averted in some cases, and no more than a couple of thousand dollars per life in many cases. Critics of tight regulation of nuclear facilities charge that it is irrational, and a genuine tragedy, that due to the American public's paranoia over nuclear power millions to billions of dollars must be spent on ever stricter, redundant, and unnecessary safety features and procedures that result in an estimated savings of very few lives, while the same money could be put to much more effective use saving lives by other means.

Questions

1. Do you agree with the idea that we should maximize the number of human lives saved with the amount of money we spend? Or should other factors also be taken into account?

2. Is safety at a nuclear reactor important for its own sake, even if it is an expensive way to "save lives"?

3. Do most people in more developed or industrialized countries put their own lives and safety before that of people in less developed countries, even if it is more expensive to protect lives in the more developed countries? Explain.

Source

Cohen, B. L., 1990, *The Nuclear Energy Option: An Alternative for the 90s.* New York and London: Plenum Press.

-8-

Differing Approaches to the Population Problem

With approximately six billion people on Earth today, many environmentalists consider global overpopulation to be *the* central environmental issue. Humanity's detrimental effects on the planet are, in most cases, the cumulative effects of many, many individuals. Each individual contributes to the waste and degradation of the planet. The assaults of a few people may be easily absorbed by the globe's natural ecosystems, but the pressure of six billion people has overwhelmed the natural defenses of Earth.

Many researchers--demographers, economists, biologists, ecologists, politicians, and others--have addressed the issue of global population over the past fifty years, yet no consensus has arisen as to whether there even truly is a global overpopulation problem. Some economists, for instance, consider population growth to be healthy--an increase in people expands markets, and therefore production and wealth. Most scholars do acknowledge that there are too many humans on Earth, however. Given that there are only finite resources on Earth, many argue, it follows that if there were fewer individuals, then the average standard of living for those who share the planet would naturally be higher. Among those who advocate stemming global population growth there are often fundamental disagreements as to how this can be done.

As Dr. Nathan Keyfitz, professor emeritus of demography at Harvard University, has pointed out, how a particular scholar approaches the subject of global overpopulation is often largely a function of his or her training. The basic dichotomy can be reduced to the difference between biologists and economists.

Keyfitz cites the following seemingly straightforward question as an example to make this point: Should a rich nation, committed to discouraging population growth around the world, send food aid to help feed the hungry in overpopulated regions of the planet? Based only on analysis through their respective disciplines (not taking ethical or moral factors into account), the biologist might easily answer this question with an emphatic NO whereas the economist would almost certainly say YES.

The biologist perhaps reasons that it takes food to keep people alive, and it takes even more food to keep people (particularly women) healthy so that they can reproduce and produce more people. Sending food to hungry, overpopulated areas therefore will simply encourage reproduction with the ultimate result that there will be more hungry people. To put it crudely, if harshly, in the short term it is better to let famine and associated disease run its natural course. The population will be reduced and those who ultimately survive will be better off. Food aid from rich nations to starving areas is self-defeating. Understandably, such an attitude is not morally, socially, or politically acceptable to most people. We tend to be trained, from an early age, to help our fellow human beings in need. Letting people starve to death seems incompatible with the ideal of helping them. In many rich nations the most politically acceptable form of foreign aid is delivering food directly to poor, hungry nations. Not only do many people view such charity as morally right, but it also serves the practical function of helping the farmers of the rich country when the national government purchases their surplus grain to donate to a needy country.

Even setting moral, ethical, and political considerations aside, the typical economist or demographer would advocate food aid to poor, overpopulated, hungry nations as a way to help control global overpopulation. The economist could argue that studies from around the world have found positive correlations between a higher standard of living, better education, access to an abundant food supply, low infant mortality rates (promoted by good nutrition) and low birth rates. Low birth rates mean lower population growth rates (if the birth rate is low enough, the size of the population may even decrease). Therefore, sending food aid to a poor, hungry nation will help that nation to increase its standard of living which will ultimately result in lower birth rates.

How can biologists and economists have such starkly different views on the same subject? Which view, if either, is correct? Keyfitz suggests that the differing views are inherent in the nature of the respective disciplines of biology and economics. Biologists are ultimately natural scientists, and like most scientists they are interested fundamentally in looking at one or a few factors at a time--the ultimate scientific tool is the controlled experiment that analyzes a single variable. Clearly, if one holds all other factors constant and simply deprives a population of adequate nutrition, then the population will ultimately decrease in numbers. This works whether one is studying bacteria in a Petri dish, catfish in an isolated aquarium, or humans in an impoverished nation with inadequate domestic agricultural output and no input from the rest of the world. In every case at least some individuals in the population will reach the point of starvation and death. Starvation continues to be a very real problem for many people in some underdeveloped countries.

But in the real world of humans in an interconnected global society, economists and demographers may counter, situations are rarely as simple as imagined by some biologists. Numerous different factors help influence the collective behavior of humans in a certain population. Starvation, famine, and disease may ultimately kill people, but this is an extreme scenario that is not typically realized. Rather poverty-stricken, poorly educated, poorly nourished persons often over-reproduce, perhaps out of ignorance, or to make up for the infant and child mortality rates in such situations. If it is for the latter reason, they often over-compensate such that the population steadily grows in size. In contrast, better-educated people who have access to good nutrition, modern medical care, and the other advantages of modern life tend to produce fewer offspring. Again, this leads to the conclusion that one way to combat population growth is to provide food aid as well as other forms of aid, such as promoting education, modern medicine, and development in general.

Questions

1. Give several examples of how the actions of a few individuals might be harmless, but when the same actions are performed by hundreds of millions or billions of people global ramifications may occur. Is global warming due to increased concentrations of greenhouse gases a case in point?

2. Of course the above analysis is extremely crude and does not represent the views of all biologists, economists, or demographers--but it can be useful in thinking about such issues to generalize the extreme positions. With this in mind when considering global overpopulation, with which group do you identify the most, the biologists or the economists/demographers? Why? Which group's arguments and reasoning do you find more convincing?

3. Biologists are natural scientists while economists are social scientists. How does this affect each group's thinking on population issues?

4. What role does religion, ethics, or morality play in your thinking about population issues? Do you find either the biological or the economics viewpoint to be ethically more acceptable? Even if you believe that the biological viewpoint is "correct," might you espouse the economic solution of food aid based on your moral and ethical values?

Source

Keyfitz, Nathan, 1994, "Demographic Discord." *The Sciences* (September/October 1994), pp. 21-27.

-9-

Famine and Overpopulation in Ethiopia

Ethiopia, in eastern sub-Saharan Africa, is a land of highlands, plateaus, and mountains, tropical lowlands and plains, and the Great Rift Valley. The country has a long and illustrious history, traditionally dating back three thousand years to the ancient Abyssinian empire once ruled by the legendary Menelik I who was said to be the son of the Biblical King Solomon and the Queen of Sheba. In recent decades, however, Ethiopia has been plagued by civil war, overpopulation, and periodic mass famines. In 1993 the northern province of Eritrea (bordering the Red Sea), which had been joined to Ethiopia after World War II, formally declared its independence. There are signs that other parts of Ethiopia may follow suit, thus fracturing the nation even further.

Currently (1996) Ethiopia has a population of 56 million and Eritrea has a population of 3.6 million. In 1950 Ethiopia and Eritrea had a combined population of only 19.6 million, but by 1990 this had grown to 49.8 million. Given that in the early 1990s the average annual population growth rate for the region was over 3%, projections are that the combined populations of Ethiopia and Eritrea could reach 130 million by 2025.

Ethiopia (including Eritrea) is primarily an agricultural nation. Over 80% of the population is involved in farming of some sort. Agricultural products account for 50% of the country's Gross National Product, and for 90% of its exports (much of this in the form of coffee). Over twenty-five years ago it was predicted that the agricultural potential of the country should be able to feed 100 million people. However, reality is very different. Ethiopia has been plagued by major famines and as of the late 1980s and early 1990s the average per capita calories available to residents was only 73% of minimum requirements--effectively meaning that most people were extremely underfed, even literally starving.

Ethiopia's woes have been attributed primarily to social and political factors, although natural fluctuations in the weather may have contributed as well. In ancient times the farmers of the Ethiopian region generally followed sustainable agricultural practices, such as rotating crops, allowing fields to periodically lie fallow, not overtaxing water supplies, and so forth. In general those farming the land owned it and benefited from the products of the land, and so they had a strong incentive to preserve its productivity. As early as the twelfth century, however, a feudal system began to arise that took control of the land and its products away from local farmers and gave it to wealthy nobility who became large landowners. Peasants had to work the land and give a percentage of the produce to the landowner. This system reached its culmination in the mid-twentieth century under the Emperor Haile Selassie (who ruled Ethiopia from 1930 to 1974). Selassie even confiscated large tracts of land to either keep for himself or give as gifts to military officers, high-ranking officials, and nobles.

The farmers working the land, but not owning the land, lost any incentive to maintain the land in a sustainable manner. When forced to increase yields, the farmers generally did so by overworking the land: planting crops repeatedly without rotating crops or allowing for fallow periods, planting marginal areas that should not have been planted, clearing forests to produce fresh fields, and allowing livestock to overgraze pastures. Farmers were also encouraged to plant cash crops, such as coffee, that the landowners could readily sell on the foreign market.

Such unsound agricultural practices quickly robbed the soils of nutrients, caused massive soil erosion, led to massive deforestation (by the 1990s the forests of Ethiopia and Eritrea had been reduced to about a quarter of what they were a hundred years earlier), and ultimately harvests could not be maintained. Fuel wood is unavailable in many parts of the region and large stretches of land are barren. According to some climatologists, the massive deforestation of the region may have reduced local humidity levels and thus contributed to the pattern of decreased rainfall observed in Ethiopia since the early 1970s.

Over time the rich got richer and the poor got poorer and more discontented. Toward the end of his reign Emperor Selassie attempted to institute agricultural reform practices that would benefit the land and peasant farmers, but he was blocked by the wealthy landowners he had helped to maintain. Furthermore, modern western farm equipment and techniques that were introduced to the area were often unsuited to the local soils, climatic conditions, and culture.

In 1972 a terrible drought hit Ethiopia. The Ethiopian government adopted an approach of silence, refusing to admit that the country was badly hurting; thus, potential foreign aid in the form of food, money, or equipment was not initially available. By 1973 millions were starving, and an estimated 8% of the population (about 2.5 million) died of malnourishment or disease associated with hunger; another 20 million (well over half the population) were perpetually short of food. Drinking water was a major concern; it was estimated that less than 3% of the population had access to uncontaminated fresh water.

The 1972-74 famine fostered extreme discontent, strikes, student demonstrations, and army mutiny which resulted in the 1974 overthrow of the imperial government and the rise of a Marxist regime headed by Lieutenant Colonel Mengistu Haile Miriam. Mengistu and his supporters controlled Ethiopia until mid-1991. During that time they did little to help the peasant farmers. In fact, at various times they formed collectivized state farms, forcefully resettled farmers from one region of the nation to another, and instituted programs of "villagization" wherein farmers were made (sometimes at gunpoint) to move to villages, ostensibly so that they could be provided with health, social, and educational services. In reality, most of the Mengistu government's actions were aimed at breaking up potential internal threats to the regime, suppressing rebellious ethnic and religious groups, and appeasing the populations of cities and the army (which might always mutiny). With the latter in mind, food prices were set at about 70% below market values, further frustrating and demoralizing the peasant farmers. As they continued to suffer, many farmers and their families attempted to escape to Somalia or the Sudan.

Food shortages constantly plagued the country, and in 1984-1985 another major famine brought on by drought and exacerbated by poor agricultural practices and government actions struck. This time at least a million people starved to death, but foreign aid in the form of hundreds of millions of dollars and hundreds of thousands of metric tons of food also poured into Ethiopia. However, much of this aid was used by the ruling government to suppress rebellious elements in the country. In the north there were particularly rebellious elements and aid was not permitted into certain territories even as the people starved to death. There were also many reports of government and military personnel confiscating food and relief supplies for their own use, abusing foreign volunteers, and searching relief camps for rebel elements.

The famine added to the discontent and strengthened rebel causes. Further famines struck the country in the late 1980s, but as before the foreign aid that poured in was often not allowed to reach the people most in need, abuses occurred, and several factions (united under the name of the Ethiopian People's Revolutionary Democratic Front [EPRDF]) continued to gain power. As a result, Mengistu's government increased its efforts to suppress the rebels--by the late 1980s some 70% of the country's economic resources were being spent on weapons to fight rebel factions--and less and less was left for the farmers or the general population. The Mengistu regime was finally defeated in 1991 and Mengistu fled to Zimbabwe. The EPRDF set up a transitional government after taking power in 1991. On May 24, 1993, Eritrea (the heart of much of the rebel activity) declared its independence and is now recognized as a separate nation. In Ethiopia a new constitution was ratified in December, 1994, and the first general elections were held in 1995.

Today Ethiopia remains a ravaged country. The population is increasing at a tremendous rate, the land is devastated, food continues to be in short supply, and the country remains at the mercy of the rains--droughts can mean devastation, and heavy downpours can further damage and wash away the denuded soils. Eritrea's secession from Ethiopia has not gone over well with many Ethiopians. Without Eritrea the remainder of the country is landlocked; having ports is critical for foreign trade and foreign aid, especially famine relief. Eritrea has stated that it will not interfere with the flow of food and goods from its ports into Ethiopia proper, but such promises may be easily broken.

Questions

1. Despite the famines, poor living conditions, and internal conflicts, why does Ethiopia's population continue to increase? How do people respond when they are not sure how many, if any, of their children will survive until adulthood? Might children provide a form of "social security" for the parents in their old age?

2. How does the history of Ethiopia exemplify the histories of many nations, especially in

the developing world? Is it understandable that the Mengistu regime spent most of its time and resources fighting the opposition rebels? What do you predict for the future of Ethiopia? Might history repeat itself once again?

3. What are the primary factors responsible for the massive famines that have swept through Ethiopia in recent years? Should foreign governments and private relief organizations send aid to the country? Or is providing relief perhaps sending the wrong message or simply supporting a corrupt regime?

4. What, if anything, should the United Nations, the United States, or other developed countries do to help the general populace of Ethiopia? Should we be interfering in the internal affairs of a sovereign nation? If so, what circumstances justify such interference?

Sources

Famighetti, Robert, editor, 1995, *The World Almanac and Book of Facts 1996*. Mahwah, New Jersey: Funk and Wagnalls Corporation.

Kaufman, Donald G., and Karla Armbuster, *Finding Our Niche: The Human Role in Healing the Earth*. New York: HarperCollins College Publishers.

World Resources Institute, 1994, *World Resources 1994-95*. New York: Oxford University Press.

Section 2

Problems of Resource Depletion

-10-

Making the Decision: Building a Highway to a Hospital or Protecting an Endangered Beetle

The American burying beetle, *Nicrophorus americanus*, is a large black and orange beetle that acquired its common name because it buries dead animals. The beetles actively search out carrion. When a carcass--perhaps a dead mouse or a baby bird of the right size and weight--becomes available, the beetles descend upon it and fight over it until a single pair, male and female, take the prize. The pair then strip the body of feathers or fur and bury it below the soil surface. The female lays her eggs on the carcass, and once hatched the parents actively feed the carcass to the larvae.

For millennia American burying beetles disposed of small animal carcasses in this manner. A hundred years ago they were found throughout middle and eastern North America. Today, however, they are extremely rare--known only from isolated populations in Oklahoma, Arkansas, and on Block Island (an island off the Rhode Island coast). In recognition of its rarity, *Nicrophorus americanus* was placed on the endangered species list by the United States Fish and Wildlife Service in 1989. Why the population of these beetles declined so rapidly, in less than a hundred years, is still unknown. Perhaps it was simply the general encroachment of humans that seriously disrupted its habitat and ancient ways.

According to the Federal Endangered Species Act, once a species is listed as endangered then it must be protected from harm by humans. This is not a problem relative to sparsely populated Block Island (where, in fact, the major beetle population lives on property owned by The Nature Conservancy), but elsewhere there have been major run-ins between the beetles and human activities. One of the most notable occurred in southeastern Oklahoma.

In this area live a number of Native Americans. The Choctaw Nation Indian Hospital, located south of the San Bois Mountains near the town of Talihina, provides medical care for more than twenty thousand local Native Americans. This hospital is conveniently located on state Highway 82--convenient, that is, if one happens to live south of the San Bois Mountains: for many years the highway ended at the mountains and began again thirteen miles further north, across the mountains. There were only two ways to get from the northern part of Highway 82 to the southern part: either follow dirt roads and trails through the mountains (which required four-wheel drive) or take country roads to make a fifty-mile detour. Since many Native Americans live north of the mountains, it is understandable that connecting the two sections of Highway 82 became a top priority of the Choctaw Nation road committee.

By 1988 the Choctaw Nation, the Oklahoma Department of Transportation, the Oklahoma legislature, and the Federal Highway Administration had all agreed to support the joining of Highway 82 through the mountains. But then *Nicrophorus americanus* entered the picture.

All of the feasible routes for the highway ran through areas occupied by populations of the American burying beetle. The Fish and Wildlife Service, charged with protecting endangered species, would not allow construction. Many people questioned this decision. It seemed that the interests of an insect were put ahead of humans.

Questions

1. In your opinion, what is more important: easy access to a hospital (serving many relatively poor Native Americans), or the preservation of a seemingly insignificant beetle species? If it could be demonstrated that a decision against extending the highway would result in a certain number of human deaths per year (due to limited hospital access), how would this affect your opinion of the situation?

2. Given the circumstances--that the beetle does fall under the auspices of the Endangered Species Act which the Fish and Wildlife Service must enforce--was there any choice but to cancel the highway? What if, unknown to scientists and other authorities, the range and population size of the beetle is actually much larger than first suspected?

3. Is a refusal to allow the highway to be built necessarily a judgement that beetles are more important than humans, as some people concluded?

4. What moral, ethical, and legal obligations do we have toward an endangered species of animal? What moral, ethical, and legal obligations do we have toward our fellow human beings?

Postscript

The Fish and Wildlife service canceled completion of Highway 82 in 1991, but reversed its decision in 1992. What changed the minds of administrators was not a rethinking of the situation described above, but continued studies that demonstrated the American burying beetles' Oklahoma range to be much more extensive than first thought. Still, along the route of the highway the Fish and Wildlife Service insisted that all American burying beetles be trapped alive and relocated to safe territory.

Sources

Mann, Charles C., and Mark L. Plummer, 1995, *Noah's Choice: The Future of Endangered Species*. New York: Alfred A. Knopf.

Raithel, C., 1991, *American Burying Beetle (Nicrophorus americanus) Recovery Plan*. Washington, D.C.: United States Fish and Wildlife Service.

-11-

The Wildlands Project

In 1991-1992 a group of conservation biologists and biodiversity activists (as they referred to themselves) founded the Wildlands Project. Leading founders of this project included Dave Foreman, also a co-founder of the radical environmentalist movement Earth First!, and Dr. Michael Soule, a prominent conservation biologist at the University of California, Santa Cruz. Allied with the Wildlands Project are numerous individuals and grass-roots organizations across North America. The goal of the project is extremely ambitious--"to help protect and restore the ecological richness and native biodiversity of North America through the establishment of a connected system of reserves." (Foreman, executive editor, 1992, p. 3)

In order to accomplish this goal, the Wildlands Project would augment existing parks, wildernesses, and refuges by developing a system of large core reserves devoted to natural wildlife and habitat, biodiversity and ecological processes. The core reserves would be linked by biological corridors to allow for the dispersal of species, genetic exchange between populations, and even the migration of organisms in response to external factors such as climate change (for example, potential global warming). Only the most minimal of human intrusions would be allowed in the core reserves and corridors (perhaps a few hikers on a limited basis). Wide buffer zones would be established around all core reserves and corridors; only limited human activity, compatible with the goals of the project and protection of the core reserves and corridors, would be allowed. All intensive human activity--agriculture, industry, municipalities, and civilization in general--would be placed well outside of the buffer zones.

The scope and magnitude of the Wildlands Project can be seen in this excerpt taken from their mission statement: "Our vision is simple: we live for the day when Grizzlies in Chihuahua have an unbroken connection to Grizzlies in Alaska; when Gray Wolf populations are continuous from New Mexico to Greenland; when vast unbroken forests and flowing plains again thrive and support pre-Columbian populations of plants and animals; when humans dwell with respect, harmony, and affection for the land; when we come to live no longer as strangers and aliens on this continent." (Foreman, executive editor, 1992, p. 3)

To complete the Wildlands Project would require decades, or possibly centuries. It would also require a change in the general mindset toward wilderness, and perhaps a major change in the very fabric of North American culture. It might also require a decrease in the human population of North America in order to accommodate all of the core reserves, corridors, and buffer zones. If carried out in full, tens of millions of people might have to be relocated; roads, buildings, and other infrastructures would be demolished and removed from certain areas; and ultimately about half of the United States might be devoted primarily to species other than humans. Whether the political and economic will to accomplish such feats can ever be organized is questioned by many, but then no one knows what the future may bring. Proponents of the Wildlands Project do point out that it would create many jobs, at least in the short-term (removing all those roads, buildings, and powerlines would require a lot of labor). Technology of the twenty-second century may make it possible for humans to live in smaller areas with less in the way of material goods, even if the goods are of higher quality and more versatility (witness the decrease in size but increase in power of computers and other electronic equipment over the last thirty years). This may make it possible to fulfill the Wildlands Project vision.

Questions

1. In the quotation from the Wildlands Project, do you perceive a bit of romanticism about nature? Even if you do, is this necessarily bad? Might it in fact be good?

2. How realistic--how feasible--do you think the Wildlands Project is? Is it worth spending time, energy, and money on? Would you support it?

3. If carried to completion, the Wildlands Project would displace many people (either voluntarily or involuntarily) in order to restore "natural" ecosystems. Would such restored ecosystems be truly natural, or simply a human reconstruction of what is thought to be natural? Would it matter either way? What about the people who would be displaced--should the lives of perhaps tens of millions of people be disrupted to restore ecosystems? Who will benefit most from such restored ecosystems?

Sources

Easterbrook, Gregg, 1995, *A Moment on the Earth*. New York, Viking.

Foreman, Dave, executive editor, 1992, "The Wildlands Project." *Wild Earth Special Issue* (P.O. Box 482, Canton, New York, 13617).

-12-

The Environment Held Hostage: The Gulf War Experience

In August 1990 Iraq overran its oil-rich neighbor Kuwait and declared Kuwait the 19th province of Iraq. This sparked an international crisis, and among other things Iraqi leader Saddam Hussein took hostage Westerners trapped in Kuwait as an attempt to prevent foreign powers from coming to the aid of Kuwait. It was soon decided that such hostage-taking was not having the desired effect, and the hostages were released by the end of 1990. Hussein then declared that he would hold the Kuwaiti environment, and ultimately the global environment, hostage. If anyone dared to interfere with his takeover of Kuwait, Hussein threatened to release oil from storage tanks onto the desert and into the Persian Gulf, to destroy oil refineries and oil wells, and to set as much oil on fire as possible. The basic idea was to cause maximal damage. Saddam Hussein hoped that his posturing would keep the nations of the world from attacking him, but his bluff was called.

The world did not give in to Hussein. The United Nations Security Council set a deadline for Iraq to withdraw from Kuwait, and once the deadline passed a multi-nation coalition, led by the United States, attacked Iraq. The first air strikes began on January 16, 1991, and a ground attack to retake Kuwait began on February 23, 1991. By the end of February the Iraqi forces had been defeated and the independence of Kuwait was restored.

True to his word, however, Hussein ordered the retreating Iraqi troops to blow up oil refineries, spill or dump oil stores, and set fire to as many oil wells as possible. The world was horrified at the senseless, wanton destruction of both the valuable oil and the fragile desert ecosystem of Kuwait. An estimated six to eight million barrels (250 to 335 million gallons, or about 25 times as much oil as spilled from the *Exxon Valdez*) was dumped into the Persian Gulf. Seven hundred oil well were set on fire, and plumes of smoke and pollutants filled the sky. Lakes of oil covered parts of the desert, and in some cases caught fire.

Initially no one was sure how much damage Hussein's ecological sabotage would cause. Some predicted that the huge influx of smoke and carbon dioxide into the environment could seriously affect local and global climates--and indeed it did. During the summer of 1991 temperatures in Kuwait were 10 to 27 degrees Fahrenheit lower than average, due to the

smoke that blocked the sunlight. Acid rain fell in the Gulf region and could be detected over a thousand miles away. But the effects were not as bad as the worst predictions, for instance that such significant cooling would reach around the globe. Fortunately
the worst-case predictions concerning how long it would take to put out the fires and cap all the oil wells also proved false. At the beginning of the disaster several different authorities thought it would take anywhere from two to five years to get the job done. As it was, all of the fires were put out by November, 1991. Still, immense devastation to the land, the people, and the flora and fauna occurred.

Clearly Hussein was directly responsible for the damage, but afterwards some critics of the war questioned whether the way the United States and the world handled the situation might not be partly to blame for the environmental devastation. Since Hussein had clearly threatened to take the actions that he did eventually order if attacked, some argued that the U.S. and U.N. took on at least partial responsibility for the destruction when the decision to launch a military attack was made. Instead, these critics argued, non-combat pressures such as trade embargoes and the like should have been utilized.

Questions

1. Do you think there are ever any political or other reasons that justify holding the environment as a hostage? In the case of the Gulf War, Hussein's threat did not stop his army from being attacked; however, he did make good on his threat. Do you think Hussein set a precedent for other nations or large terrorist organizations?

2. If a person owns a piece of pristine, old-growth forest and threatens to burn or clear-cut it unless the government or an environmental organization is willing to purchase it at a greatly inflated price, could this be considered a type of environmental blackmail or hostage-taking? In your consideration of such an incident, does it matter to you whether such tactics are legal or illegal? That is, would you condemn such actions only if they were illegal?

3. Do you think those critics who initially predicted "worst-case scenarios" of the consequences of Iraq's policy of environmental devastation had an agenda of their own that might influence their predictions? If so, what might that agenda have been? (Remember, Republican George Bush was then the U.S. president and already preliminary plans were being made for the 1992 election year.)

4. Do you think that the United States and United Nations should share any of the blame for the destruction caused by Iraq's retreating army? Justify your answer.

Sources

Buzzworm Magazine Editors, 1992, *1992 Earth Journal: Environmental Almanac and Resource Directory*. Boulder, Colorado: Buzzworm Books.

Newton, Lisa H., and Catherine K. Dillingham, 1994, *Watersheds: Classic Cases in Environmental Ethics*. Belmont, California: Wadsworth Publishing Company.

-13-

The *Exxon Valdez* Oil Spill

In the early morning hours of March 24, 1989 the tanker *Exxon Valdez*, carrying some 60 million gallons of crude oil from the Prudhoe Bay area of Alaska, ran aground on Bligh Reef in Prince William Sound. The Alaskan pipeline carries crude oil from the North Slope oil fields of Alaska near Prudhoe Bay south across the state to a tanker terminal just south of Valdez, Alaska. At the terminal tankers load the oil, then make their way out of Prince William Sound into the Gulf of Alaska and the open ocean.

The *Exxon Valdez*, a 987-foot oil tanker owned by Exxon Shipping Company (part of the larger Exxon Corporation), left the port at 9:30 p.m. on March 23, 1989, and within three hours had run aground, spilling 10.8 million gallons of oil into the sound. The oil quickly spread over the water; after ten days the oil slick covered over a thousand square miles. Ultimately over 12,000 cleanup workers descended on the area, and Exxon spent approximately $2.2 billion on the cleanup operations. The spill took a heavy toll on the wildlife of the area. Although estimates of the damage vary widely, at least tens of thousands of sea birds died, over a thousand sea otters and several hundred harbor seals died, and unknown numbers of fish and invertebrates succumbed to the oil. It was estimated that despite the extensive cleanup efforts, only 5 to 9% of the oil was actually recovered. Another 20 to 40% of the oil evaporated; the rest washed up on the beaches or sank to the ocean floor, eventually to be decomposed naturally.

Over $2 billion in cleanup costs were not the only expenses incurred by Exxon. The person in command of the *Exxon Valdez*, Captain Joseph Hazelwood, was apparently drunk at the time of the accident. Furthermore, the crew in general was badly fatigued and overworked. At the very moment of the accident the ship was being piloted by a third mate who was not licensed to navigate through Prince William Sound, rather than by the captain himself. As a result of findings by the National Transportation Safety Board and a federal jury that the actions of that night were negligent and reckless, Exxon was required to pay over $1 billion in fines to the federal government and the state of Alaska, and thus far several billion dollars worth of damages have been awarded to fishermen, property owners, and others bringing claims against Exxon. Hundreds of claims are still pending.

Such are the bare facts of the *Exxon Valdez* oil spill, but many questions remain unanswered. An immediate question is: How could such an accident occur? The easiest answer, at least for many people, is to blame it on the man in charge, Captain Joseph Hazelwood. Hazelwood was clearly intoxicated at the time of the accident. According to later testimony, he had had fourteen shots of vodka on the afternoon of March 23, 1989. Witnesses testified that they had smelled alcohol on Hazelwood's breath immediately before and after the accident; nine hours after the accident blood tests on Hazelwood showed that he still registered a 0.061% alcohol blood level.

Exxon officials were certainly aware of Hazelwood's past alcohol problems. He had been twice convicted of drunk-driving charges, and at the time of the *Exxon Valdez* accident his automobile driver's license had been suspended. In 1985, with the approval of Exxon officials, Hazelwood had participated in a twenty-eight day alcohol rehabilitation program. Exxon has a strict policy that prohibits alcohol consumption on its tankers. Still, witnesses testified that a month before the accident Hazelwood was observed to consume alcohol in the ship's lounge with his officers.

But it was not just the problem of Hazelwood's drinking that may have led to the accident. Exxon, in order to save money, had cut back on the size of its tanker crews. At the time of the accident the ship had only twenty crew members; effectively this meant that each crew member had to put in an extra four or five hours per day (they were working twelve and thirteen hours a day). As a result, the crew was badly fatigued, a situation conducive to mistakes occurring. At the time of the accident the third mate was off course, he had missed a signal light, and he was attempting to avoid ice in the sound when he hit the reef.

Once the ship hit rock, the crew did act quickly and responsibly to seal off compartments in the ship, keeping it from spilling even more oil or sinking. The accident would not have been nearly as bad, however, if the *Exxon Valdez* had been equipped with a double hull. Converting the ship from single to double hulled would have cost an estimated $20 million to $30 million, but in hindsight could have saved Exxon billions of dollars. In the early 1970s government officials had suggested that all oil tankers plying Alaskan waters be double hulled, but due to lobbying from the oil industry, such a requirement was never put into place.

After the accident occurred, there were many problems surrounding the cleanup operations. Alyeska Pipeline Service Company (a consortium of seven oil companies working in Alaska) and Exxon responded very slowly and with inadequate equipment and manpower, according to many critics. Cost-cutting and reduction in personnel on the part of Alyeska over the years meant that the equipment that was supposed to be available for cleanups was not. Cleanup procedures became bogged down in bureaucracy and red tape. Much of the money spent on cleanup was, according to some critics, wasted. A prime example was $18 million spent to rescue and treat oil-soaked otters. In total, 357 otters were treated but only 197 survived to be returned to the wild. Of those that survived, it was feared that many might be suffering from a type of pneumonia and not live long at any rate, and might even spread the disease to healthy otter populations. $18 million spent to "save" only 197 meant that each otter successfully treated cost over $91,000.

The *Exxon Valdez* accident was certainly unfortunate, but some critics suggest that the cleanup operations may have been even more damaging to the Prince William Sound ecosystem than the oil spill itself. Immediately after the spill the most pessimistic commentators decried the total loss of the sound; the disaster had destroyed forever the pristine Prince William Sound. But perhaps the real disaster was the hundreds of boats and ships, the thousands of workers, the airplanes and helicopters that descended en masse on the sound--bringing their own forms of pollution, such as leaking oil, noise, and exhaust fumes. The base camps on the shores, the trampling of the shores, and the physical crushing of the intertidal ecosystems under the weight of cleanup equipment all added to the destruction. Many of the methods used to address the oil spill were primarily cosmetic. In places rocks and stones were scrubbed by hand, even using small hand brushes and toothpicks. Hot water under high pressure was used to wash away the oil from beaches. But such drastic measures, especially the hot water, killed all life, including the microbes so important in decomposing dead organic matter and serving as a base of the food chain. The "cleaned" beaches were left sterile and dead. In contrast, uncleaned areas of the sound (left as controls to see what effects the oil spill would have) fared much better. After several years they had effectively cleansed themselves through microbial activity, wave action, and the natural degradation of the crude oil.

One critic of the cleanup operations after the spill contended that the best possible course of action would have been simply to contain the spill as well as possible (so as not to allow further oil to leak), then do nothing. Over time the sound would naturally clean and heal itself. The two billion dollars wasted on cleanup efforts could have been better spent on energy conservation and efficiency measures (so as to decrease our need for oil), the development of alternative energy sources, the upgrading of oil tankers (perhaps installing safer double hull systems), or any of a number of other worthy causes.

Politically and socially, however, one could argue that cleanup operations had to be undertaken. The *Exxon Valdez* oil spill was well-publicized and the public demanded that action be taken. Exxon, as well as state and federal authorities, suffered an inordinate amount of bad publicity due to the spill itself. To do nothing, even if such inaction would have ultimately been the best course of action, would be seen as inexcusable. If nothing else, public relations demanded a massive cleanup operation.

Questions

1. Do you think that Joseph Hazelwood is truly to blame for the accident, or is he simply a convenient scapegoat? Are the policies and attitudes of the Exxon Corporation and its top officials the true culprits? Could Exxon officials have believed that despite his past drinking problems, Hazelwood was rehabilitated?

2. Should Alyeska and Exxon have been better prepared for an accident on the magnitude of the *Exxon Valdez* oil spill? Alyeska had estimated that such a large spill would happen only once every 241 years. Do you think this estimate was faulty? (Consider that the *Exxon Valdez*, although significant, is not even the largest tanker oil spill. For example, a collision between two tankers off Trinidad and Tobago on July 19, 1979, spilled some 88 million gallons of oil; the grounding of a tanker near Portsall, France, on March 16, 1978, spilled

65 million gallons; a collision in Galveston Bay, Texas, on November 1, 1979, spilled 10.7 million gallons; and a grounding off the Shetland Islands on January 5, 1993, spilled 26 million gallons. The list of significant oil spills could be multiplied by a factor of ten or more; one can argue that oil spills are simply an expected part of the petroleum industry.)

3. Do you think the money used for cleanup operations after the Exxon Valdez spill was well-spent or wasted? Exxon produced impressive before and after pictures of some beaches in order to demonstrate the success of their cleanup, but critics argued that the cosmetically restored beaches were biologically dead. On what basis must one judge the value of cleanup operations? Would inaction have been a politically feasible course of action?

Sources

Easterbrook, Gregg, 1995, *A Moment on the Earth: The Coming Age of Environmental Optimism*. New York: Viking.

Famighetti, Robert, editor, 1995, *The World Almanac and Book of Facts 1996*. Mahwah, New Jersey: Funk and Wagnalls Corporation.

Jennings, Marianne Moody, 1993, *Case Studies in Business Ethics*. Minneapolis/St. Paul: West Publishing Company.

Newton, Lisa H., and Catherine K. Dillingham, 1994, *Watersheds: Classic Cases in Environmental Ethics*. Belmont, California: Wadsworth Publishing Company.

-14-

The Dangers of Conservation

Professor Bernard L. Cohen, in his book advocating nuclear power (1990), argues that while energy conservation and efficiency improvements may decrease our need for energy, such measures can also be very dangerous. This concept he quantifies using the LLE (Loss of Life Expectancy) statistic. To give a few examples of what he is referring to, if we use smaller cars to save energy we may increase our LLE by up to 60 days (smaller cars are currently more dangerous than larger cars). Doubling the amount of bicycle riding increases the LLE by 10 days for the average American. Superinsulating and sealing buildings increases our exposure to radon and indoor air pollution. Reducing the amount of lighting used inside buildings would increase the number of falls and other accidents. Decreasing the number of outside street lights would increase the number of night-time motor vehicle accidents, and also increase the number of robberies, burglaries, and muggings. Thus energy conservation, it is argued, is detrimental to human life.

There is a simple counter-argument to the above train of thought. There are many ways of increasing energy efficiency without sacrificing comfort or health. Compact fluorescent lights can generate just as much usable light with much less energy, so there should be no increase in accidents. Special bike paths can be designed and built so that it will actually be safer to ride a bike under some situations than to drive a car. Smaller cars can be designed specifically with safety in mind--small does not necessarily need to mean dangerous. Proper ventilation systems can be installed in superinsulated buildings, and so forth.

Cohen (1990, p. 136-137) makes a more insidious assertion, playing on the fears of many people: "An important danger in overzealous energy conservation is that it may reduce our wealth by suppressing economic growth. Just to keep up with our increasing population without increasing unemployment, we must provide over a million new jobs per year for the foreseeable future. We have been succeeding in keeping up for the past several years, but that requires increasing supplies of electricity. . . . Typically, life expectancy is 75 years in economically advanced nations versus 45 years in poor nations, a difference of 30 years. . . . The greatest potential risk of overzealous energy conservation is that it may lead us down the thorny path toward becoming a poorer nation." According to Cohen (1990, p. 125) anything that increases the unemployment rate in America, such as lack of energy use, can have the following effects: "The estimated effects of a 1% increase [in the unemployment rate of the United States] for one year are 37,000 deaths, including 20,000 due to cardiovascular failure, 500 due to alcohol-related cirrhosis of the liver, 900 suicides, and 650 homicides. In addition to the deaths, there are 4,200 admissions to mental hospitals and 3,300 admissions to state prisons."

Without questioning the accuracy of Cohen's statistics, there is reason to question whether there is (or need be) a direct correlation between energy consumption, economic viability, life expectancy, and unemployment in any nation. A nation can continue to grow economically while decreasing its energy consumption. Indeed, development of a sustainable economy will in all likelihood involve more labor-intensive activities, which will mean more jobs rather than fewer. No longer will the economy be fueled by machines, but by the work of human hands.

Questions

1. Do you find the Loss of Life Expectancy statistic useful? Or is it perhaps misleading?

2. Are you convinced by the arguments claiming that conservation can be detrimental to human health?

3. Critically evaluate Cohen's contention that "overzealous energy conservation . . . may reduce our wealth by suppressing economic growth."

Source

Cohen, B. L., 1990, *The Nuclear Energy Option: An Alternative for the 90s*. New York and London: Plenum Press.

-15-

Circumventing Nuclear Waste Disposal By Reclassification

More and more nuclear waste is being generated every day, not just by nuclear power plants, but also by hospitals, laboratories, industry, and even private homes (the typical home smoke detector contains americium-241 or some other radioactive material). Much of this material is "low-level" radioactive waste, but still very dangerous. Until 1990 much of this material needed to be disposed of in a special manner--it was not to be simply thrown away in the local garbage dump. But in that year the Nuclear Regulatory Commission (NRC) decided to reclassify certain low-level radioactive wastes as below regulatory concern. What this means in effect is that now this material can be thrown away, incinerated, buried in a landfill, or whatever, without taking any special precautions. In fact, it can even be recycled into other consumer goods--perhaps goods that do not require, and normally would not contain, traces of radioactive elements in their makeup. On paper at least, the NRC has decreased the radioactive waste disposal problem.

The NRC justifies its actions by stating that under this new policy it will be able to "focus on the regulation of materials that pose much more significant risks to the public" (quoted in Galperin, 1992, p. 68). Critics of the NRC charge that the new policy will certainly allow even more human-made radioactivity to escape into the environment, and will set a precedent for reclassification of other, even more dangerous radioactive wastes. Critics are also disturbed by the NRC's current policy of sometimes granting single-case exemptions for the disposal of more dangerous nuclear wastes. In this manner a company can legally circumvent nuclear waste disposal regulations.

Questions

1. Do you agree with the decision to reclassify certain low-level radioactive wastes? Why or why not?

2. What is the underlying rationale for such reclassifications?

3. Do you believe that single-case exemptions should ever be granted for the disposal of nuclear waste? How does nuclear waste differ from other types of waste?

Source

Galperin, A. L., 1992, *Nuclear Energy, Nuclear Waste.* New York and Philadelphia: Chelsea House Publishers.

-16-

So Just How Dangerous is Plutonium?

Some antinuclear activists have asserted that plutonium is one of the most toxic substances known to man and that a single pound of plutonium could kill 8 billion people. Certainly plutonium-239, a highly radioactive, fissionable substance with a half-life of 240,000 years, is an extremely dangerous substance, but pro-nuclear advocates argue that some of the claims for the extreme toxicity of plutonium are grossly exaggerated. Dr. Bernard L. Cohen, for instance, counters that a pound of plutonium, if transformed into dust and inhaled to produce maximum effects, would produce ultimately fatal cancers in approximately two million people. This means, according to Cohen, that the toxicity of plutonium is comparable to that of nerve gas--with one difference. Nerve gas, as the name implies, is a gas that can easily be inhaled. Plutonium, on the other hand, is a heavy solid that is difficult to disperse in a particulate form so that it would be inhaled (and thus cause the maximum amount of damage). In fact, Cohen has calculated that a pound of plutonium dispersed as dust over a large city would cause only twenty-seven deaths on average; only approximately one part in 100,000 would actually be inhaled by people. Pound per pound certain biological agents, such as botulism toxin or anthrax spores, are much more toxic than plutonium.

Although it is potentially possible that inhalation of as little as a single particle of plutonium could cause cancer, the probability of this happening is extremely low. To make this point, Cohen offered to publicly "inhale 1,000 particles of plutonium of any size that can be suspended in air" or "to eat as much plutonium as any prominent nuclear critic will eat or drink caffeine." (Caffeine in pure, concentrated form is a very powerful and dangerous substance.) Such an offer may seem absurd, but Dr. Cohen justified his potential actions by saying: "My offers were such as to give me a risk equivalent to that faced by an American soldier in World War II, according to my calculations of plutonium toxicity which followed all generally accepted procedures. . . . I feel that I am engaged in a battle for my country's future, and hence should be willing to take as much risk as other soldiers" (Cohen, 1990, p. 251).

It should also be noted that plutonium is commonly used in nuclear bombs, a number of which have been detonated in the atmosphere as tests. When the typical plutonium-based bomb explodes only about 20% of the plutonium is consumed, while the rest is dispersed as fine dust in the atmosphere. According to Cohen some 10,000 pounds of plutonium have been released into the atmosphere in this manner. On the one hand, worldwide this has probably killed at least a few thousand innocent victims by the most conservative estimates. A 1991 study commissioned by the International Physicians for the Prevention of Nuclear War concluded that the fallout from atomic bomb testing will eventually cause approximately 2.4 million cancer deaths (of course, as many people now die of cancer, it is virtually impossible to know which cancer victims developed their cancers due to atmospheric plutonium dust). On the other hand, all of this plutonium dust has not wiped out humanity, as certain extreme antinuclear activists might imply it should have.

As a final point, Cohen calculates that if the United States relied on breeder reactors for its electricity needs, the country would need about 400 breeder reactors each producing about 500 kg of plutonium per year for a total of 200,000 kgs of plutonium per year. This would be enough plutonium to kill half a trillion people (5×10^{11}) if inhaled under "ideal" conditions (a virtual impossibility, as described above). This massive production of plutonium may seem like an unprecedented amount of poisonous material, but Cohen points out that every year the United States already produces enough chlorine gas to kill 400 trillion people, enough phosgene to kill 18 trillion people, enough hydrogen cyanide to kill 6 trillion people, and enough ammonia to kill 6 trillion people--to name only a few mass-produced poisonous chemicals.

Questions

1. Are you now convinced that plutonium is not quite as dangerous as you may have thought? Critically evaluate the comments by Dr. Cohen. Is it fair to compare plutonium production to the production of other types of poisonous substances?

2. What do you think of Dr. Cohen's offer to inhale or ingest plutonium? Is this just a publicity stunt?

3. Even if plutonium is not as dangerous as some people believe, is it ethically or morally right to expose virtually all people to low levels of plutonium?

Source

Cohen, B. L., 1990, *The Nuclear Energy Option: An Alternative for the 90s*. New York and London: Plenum Press.

-17-

How Easy is It to Build a Nuclear Bomb?

At present only a few countries are known to have nuclear weapons: the United States, the Commonwealth of Independent States (particularly Russia, which inherited the former Soviet Union's nuclear arsenal), Great Britain, France, the People's Republic of China, and India. (Israel possibly has nuclear capabilities.) One objection to nuclear reactors in general, and breeder reactors in particular, is a concern about nuclear proliferation. Nuclear power technology involves fissionable material, and such material in concentrated form is the fuel for a fission bomb. It is argued that fissionable material from a commercial nuclear power plant could be diverted for destructive purposes. What if a critical mass of fissionable material fell into the wrong hands, such as a hostile nation or a terrorist organization? Couldn't they then easily build a nuclear bomb?

The answer, according to most experts in the field, is no. To begin with, the fissionable material used and made in commercial power reactors is not of the proper grade to make a good fission bomb. The production of "weapons-grade" fissionable material (plutonium or uranium) from "reactor-grade" material would require either reprocessing of the material or isotope separation. For this reason, the United States government has discouraged the development of commercial reprocessing facilities, both in America and abroad. As a result there are no such facilities in the United States, but they are essential to some other countries' nuclear power programs. Reprocessing is a necessity if breeder reactors are to be used in the future. It has been asserted that hindering or banning commercial reprocessing plants would do little to hinder a group or nation that had spent nuclear fuel and wished to reprocess it solely for the purpose of building a bomb. Reprocessing technology is fully described in the published literature; no secrets are involved. In the middle 1970s it was estimated that five knowledgeable people with about $100,000 to spend (with inflation, the price might be several times this today) could set up and operate a crude reprocessing facility with the capacity to produce the raw materials for several nuclear bombs.

If a non-nuclear nation, or even a terrorist group, wanted to build a nuclear weapon there might be even easier ways to obtain the fissionable uranium or plutonium needed. One method would be to steal the material from an established weapons facility of the United States or another nuclear nation (or for that matter, steal an entire nuclear bomb or warhead). A still easier way would be to simply set up a small nuclear reactor for the express purpose of producing fissionable plutonium used in building bombs. Such a reactor would not be designed to generate electricity; it could use natural (unenriched) uranium, and it could be much smaller than a power-generating reactor and accordingly much easier, quicker, and cheaper to design and build. A bomb-making reactor could be relatively easily built clandestinely; it is in small reactors like this that most of the plutonium used for bombs by the military nuclear powers has been produced. Another option would be to divert a research reactor (used for academic, medicinal, and other purposes) to the production of plutonium for a bomb. At least 45 nations have research reactors world-wide.

Once a nation or terrorist group had about 10 kg of weapons-grade plutonium, wouldn't it be a simple thing to build a nuclear bomb? Isn't the most important and difficult thing obtaining the concentrated fissionable material in the first place? Most experts feel that even if a group has the necessary fissionable material, designing and building a workable nuclear bomb would be extremely difficult. There have been claims that bright undergraduate physics students have designed nuclear bombs on paper, but these claims have been exaggerated. According to bomb experts, what have been produced are crude sketches essentially showing how such a bomb works, not usable designs or blueprints that could be followed to build a workable bomb. To design and build a nuclear bomb requires not only considerable expertise in nuclear physics, engineering, metallurgy, and chemistry, but also experience in handling conventional chemical high explosives. In order to detonate the fissionable plutonium or uranium, chemical explosives must be set off that cause the fissionable material to implode. The timing of the explosions around the fissionable material must be critically synchronized to a small fraction (about a millionth) of a second and the neutron flux of the chain reaction must be confined within narrow limits to maximize the explosion. Because of the chemical explosives and detonators required, simply building a nuclear bomb (even from perfect blueprints) would be extremely difficult and dangerous. Certainly there are teams of experienced people who can build nuclear weapons, as evidenced by the thousands of such weapons already in existence, but to convince such a group of people to build a workable bomb for a terrorist organization is another matter.

When it comes right down to it, all the concern over nuclear proliferation may be a red herring. Any sovereign nation that wants to build nuclear weapons, and has the resources to pursue such a program, could most likely do so (given that their facilities are not destroyed by a hostile power before they can succeed, as when the Israelis bombed an Iraqi reactor that could have produced weapons-grade material). As far as covert terrorist groups are concerned, realistically they probably have little need for nuclear weapons, except perhaps as a psychological ploy. Even a small nuclear bomb would be heavy and difficult to handle. While it could be used to blow up a large building or a portion of a downtown area, killing tens of thousands of people, there are much easier ways to accomplish the same "task." Conventional explosives used to destroy the supports of a large, crowded building or stadium--causing it to collapse--could kill tens of thousands of people at one shot. Poisoning the water supply of a city, or releasing poisonous gases into the ventilation systems of large buildings, could kill innumerable people; many large office buildings are built such that the few exterior windows cannot even be opened. Conventional explosives could be used to destroy a large dam where the sudden release of the water might kill over a hundred thousand people and do billions of dollars worth of damage. Using chemical or biological agents of warfare, all kinds of creative ways of causing death and destruction could be imagined. The point is that for most terrorist groups, building a nuclear bomb would be more trouble than it was worth given all the other options to induce mayhem open to them.

Questions

1. Do you agree with the United States government position that commercial nuclear reprocessing facilities should be discouraged in order to limit potential nuclear proliferation? Why or why not?

2. By limiting nuclear fuel reprocessing is the United States dimming its energy future? Should the United States invest in breeder reactors and a full-scale reprocessing program?

3. Do you think there is any significant danger that a terrorist organization could build an operable nuclear bomb? Would they want to? Are there better, easier ways by which such a group could cause massive destruction or death?

Source

Cohen, B. L., 1990, *The Nuclear Energy Option: An Alternative for the 90s.* New York and London: Plenum Press.

Jagger, J., 1991, *The Nuclear Lion.* New York: Plenum Press.

-18-

A Uranium Plant Disguised

For years many residents of Fernald (in Hamilton County), Ohio assumed that the factory marked "Feed Materials Production Center" was simply an animal food factory. Then in the early 1980s the truth became widely known throughout the community--the facility was actually a Department of Energy (DOE) uranium plant producing nuclear weapons components, fuel rods for nuclear power plants, and tons upon tons of radioactive waste. Lisa Crawford lived across the street from that plant and became greatly concerned. In 1984 she learned that her family's well water had been contaminated with uranium and other radioactive substances. Furthermore, the DOE had known about the contamination since 1981 but said nothing until officials were forced to admit to the contamination by pressure applied by a congressman, Representative Thomas Luken of Ohio.

Lisa Crawford joined a newly formed group known as FRESH (Fernald Residents for Environmental Safety and Health) and by 1985 was the leader of the group. As she dug deeper into the problem she found more and more problems. Since opening in 1951, over 100 tons of radioactive materials had been released into the air, another 74 tons had been released into a nearby river, and 337 tons could not be accounted for at all. Furthermore it seemed clear that the radioactive releases were affecting the local residents. Rates of bladder, liver, lung, and colon cancer were all exceptionally high in Hamilton County. Lisa Crawford, her husband Ken Crawford, and eighteen other plaintiffs filed a class-action suit on behalf of approximately 14,000 citizens living within a five-mile radius of the plant. They demanded $300 million from National Lead of Ohio, the actual operator of the plant (under a DOE contract).

In 1989 the plaintiffs settled for $78 million to be paid to the residents of Fernald--with a portion of the money being designated for a fund to provide medical monitoring and care for those living near the plant. As it turned out, however, it was not National Lead of Ohio that ultimately paid: rather, based on the contract that National Lead had with the DOE, the U.S. Treasury, and thus ultimately the U.S. taxpayers, had to foot the bill.

Questions

1. Why would the Department of Energy want to disguise its uranium plant? Wouldn't it be logical that the truth would come out eventually? Or do you think they never meant to intentionally mislead the public?

2. Was it morally and ethically right for the DOE to allow radioactive materials to be released into the environment around the plant? Was it right for them to cover this up? Which was worse, the releases or the silence of the DOE?

3. What potential risks did Lisa Crawford take in speaking out against the uranium plant? What might have happened if their lawsuit had been unsuccessful?

4. Do you think it is fair that ultimately the United States taxpayers have to pay the $78 million settlement? Shouldn't National Lead of Ohio be responsible for part or all of this bill?

Source

Caplan, Ruth, and the staff of Environmental Action, 1990, *Our Earth, Ourselves.* New York: Bantam Books.

-19-

Future Water Wars

Over a billion people, over one-sixth of the world's population, lack clean drinking water, and 1.7 billion do not have access to adequate sanitation facilities, according to United Nations statistics. Among developing countries, 80% of all cases of disease are attributed to unclean water, and an estimated ten million people die as a result each year. Around the world some 80 countries and 40% of the world's population experiences some degree of freshwater shortages, and the situation does not seem to be improving. Over the next 15 to 20 years the global freshwater supply could become so badly depleted that many nations, especially in arid regions (including the Middle East and parts of Africa) could enter into armed conflict specifically over water and water rights--or so predicted Wally N'Dow, secretary general of the United Nations Conference on Human Settlements held in Beijing, China, in March 1996. Ultimately a global war over access to freshwater could ensue.

A large part of the problem is that in many areas major water basins are shared by two or more countries. Currently there are over 2,000 treaties worldwide dealing with water basins, river flows, and water rights, but in many areas the treaties are inadequate. To cite one example, the Nile River in Africa is the longest river in the world and the basin of the Nile covers parts of nine different countries, including Egypt, Sudan, and Ethiopia. Already projects to enhance the flow of parts of the Nile system by bypassing areas in southern Sudan have been contributing factors to civil war in the region. Egypt is dependent on the Nile for water. A 1959 agreement between Egypt and Sudan allocating the Nile's downstream flow between the two countries ignores the needs and demands of upstream users such as Ethiopia, where much of the water forming the Nile originates. Upstream users of rivers often claim rights to water that either originates in or flows through their territories, including the rights of use, storage, pollution, and diversion. As Ethiopia develops its irrigation and hydroelectric power potential, this could possibly reduce the flow of water into the Nile system--a situation potentially intolerable to Egypt.

Many other parts of the world have the potential for imminent armed conflict over freshwater supplies. In the Middle East the Jordan River basin is shared by Syria, Jordan, Lebanon, and Israel. The 1967 Arab-Israeli war was in part sparked by disagreements over water rights, and the potential for further conflict over water use and rights remains. Similarly, Iraq is dependent on the Tigris and Euphrates for water, but these rivers originate primarily in Turkey and Syria. Dam construction projects currently taking place in Turkey could significantly curtail water flow to Iraq and Syria, a situation that will only serve to increase tensions in the area.

In an interview held at the end of the International Conference on Managing Water Resources for Large Cities and Towns (part of the Conference on Human Settlements), N'Dow was quoted as saying: "We are silently but surely heading toward what can be described as 'water shock'. . . Without improved water resource availability and access, . . . human development will be arrested completely. I don't think societies will survive as we know them today."

A potential solution to the problem of limited freshwater supplies, some people suggest, is continued advances in seawater desalination technology. The cost of desalination has been steadily decreasing in recent years (in 1996 some processes cost less than fifty cents per cubic meter of water). If freshwater could be produced efficiently and inexpensively from abundant seawater, many of the world's most serious water problems would be solved.

Questions

1. How serious a problem do you think freshwater shortages are? Has your community ever experienced major water shortages? Is your community typical of the world situation as a whole?

2. Do you think the suggestion that future global conflicts might occur over water rights is simply alarmist exaggeration? What is to be gained by making such predictions?

3. How might the planet solve its global water problems? What are some of the main consumers of water? (For instance, agriculture, industry, power utilities.) Could less water be used? Might new technologies that can inexpensively and efficiently desalinate seawater solve our problems? Would we be ill advised to count on such a "technofix?"

Sources

Deutsche Presse Agentur, 1996, "Conference Touts Desalination." *The Boston Globe* (March 31, 1996). p. 31.

Postel, Sandra, 1992, *Last Oasis: Facing Water Scarcity*. New York: W. W. Norton.

Reuters, 1996, "World at War Over Water is Possible, Official Says." *The Boston Globe* (March 22, 1996), p. 14.

World Resources Institute, 1994, *World Resources 1994-95*. New York: Oxford University Press.

-20-

Unwanted Species in Hawaii

Isolated by vast expanses of ocean, the indigenous flora and fauna of the Hawaiian islands includes (or included) many unique species. With the coming of humans to the islands about fifteen hundred years ago, the native biota began to suffer. By 1500 at least thirty-nine native Hawaiian birds had been driven to extinction, and problems have accelerated with the invasion of the island by modern civilization in the eighteenth through twentieth centuries. Today Hawaii is home to more endangered or threatened animals and plants than any other state.

There are several factors contributing to the decline of native species in Hawaii. Historically, some species have succumbed to direct hunting by humans--this may be true of many of the Hawaiian birds that went extinct before modern man entered the islands. Other species have become progressively restricted as their habitats have been encroached upon by humans. Another important factor is the spread of species imported, either purposefully or inadvertently, to the islands by humans. Such invading species range from the microscopic to the macroscopic, from diseases to deer, and the indigenous species often have no good defenses against the alien invaders. For millennia the native plants and animals of Hawaii evolved in the absence of major predators and in isolation from many diseases found in other parts of the world. For example, Hawaiian birds lacked a natural immunity to avian malaria. With the introduction of the disease in recent times thousands of birds were imperiled.

Valiant efforts are currently being made to preserve Hawaii's vanishing species, such as the 'alala (also known as the Hawaiian crow). As large as a bald eagle, only about thirty of the birds exist at present. But as is true the world over, it is not enough to save isolated species; indeed, to make such attempts can often prove futile. The very habitat, the entire biota and ecosystem, must be preserved. In Hawaii this means controlling or eliminating invading species that disrupt the natural habitat.

Such a task is daunting; literally thousands of plant and animal species have invaded Hawaii, and every day there is the possibility of more invaders being introduced as travellers enter the islands. The state's Department of Agriculture has its hands full monitoring baggage moving through airports. Illegal non-native fruits, vegetables, or animals must be confiscated and disposed of. Tiny insects, seeds, and microbes may be the worst threats since they can easily slip by--often even unknown to the human carrier--without detection. In the spring of 1995 tiny biting insects known as nono flies were almost introduced by mistake to the islands from Polynesia, during a modern recreation of the ancient Polynesian ocean-going techniques that first brought humans to Hawaii. Fortunately the flies were discovered before the boat reached Hawaii; apparently they had been carried in fruits taken from the Marquesas Islands.

Major efforts are also being made to eradicate at least some of the invaders that have appropriated large parts of the islands. Deliberately introduced wild boars and the Axis deer, descended from seven deer that were brought from India as a gift to the Hawaiian monarch in 1867, are destroying the natural vegetation. Among plants, such species as the strawberry guava and banana poka are replacing native species, but perhaps the primary threat currently is miconia. Miconia are trees that were introduced to the islands from Central and South America by landscapers and botanical gardens. Known as "velvet trees" due to the purple colored undersides of their large leaves, they are fast growing and well-adapted to the Hawaiian climate. They can easily grow to heights of forty feet, and in the Hawaiian forest quickly outgrow, overshadow, and squeeze out the native species. They block the light for plants below them, and they also damage the water supply by shading and thus destroying mosses that naturally keep rain water in place rather than allowing it to run off the surface. Miconia have relatively shallow roots; as they crowd out deeper-rooted native species, local soil erosion often increases. A large miconia tree can produce tens of millions of its tiny seeds each year.

In attempts to control the miconia, paid workers and volunteers have been searching the Hawaiian forests, chopping down the mature trees and pulling up seedlings. In just one area, Helani Gardens, Hana, on Maui's eastern coast, 15,000 miconia were destroyed in this manner. Officials believed they had eradicated the miconia from the area, but then they found still more trees in the local forest. Around the state the trees have been sprayed with herbicides from helicopters or injected with poisons. Currently the battle continues and it is unclear who will win out--humans or miconia.

Questions

1. How are the problems faced by the Hawaiian islands typical of those faced by many islands and isolated regions?

2. If the nono flies had reached the Hawaiian islands and proliferated, what ramifications do you think it might have had for the local tourist industry? For the economy?

3. One potential way to control miconia is to import insects from Central and South America that will feed on the trees. Do you think it would be safe to release such insects into the forests of Hawaii? What if they attack not only miconia, but other plants as well? Can we ever predict exactly what the effects of introducing an alien species will be?

4. Do you think that the time, energy, and money being put into attempting to protect Hawaii's native species is worthwhile? Justify your answer.

Source

Kong, Dolores, 1995, "Hawaii's War Against the Aliens." *The Boston Globe*. October 16, 1995, pp. 29, 31.

-21-

Killing Elephants in an Overcrowded Park

Elephant populations have been on the decline for decades in many parts of Africa. Kenya and Zambia, for example, had an estimated 100,000 and 120,000 elephants respectively in 1975, but by 1993 the elephant populations were only about 18,000 in each country. Tanzania had approximately 204,000 elephants in 1981, but only 44,000 in 1993. In the Sudan elephants declined from 135,000 in 1981 to only a few thousand in the early 1990s. These dramatic declines have been attributed in part to the pressure put on elephants as the human population expands into the natural territory of the elephants, but by far the worst and most immediate cause of elephant declines has been illegal poaching.

Many African countries have taken an aggressive stance against poachers and have made valiant efforts to protect their elephant populations. However, for over thirty years a different approach has been taken in some parts of southern Africa. In parts of Botswana, Nambia, Zimbabwe, and South Africa culling of herds (the selective killing of elephants by park authorities) is carried out annually in order to artificially control the population. For instance, every year since 1967 several hundred elephants have been purposefully killed in Kruger National Park, South Africa. This practice has come under fire from many environmentalists. The basic argument against culling comes down to this: since elephant populations are generally declining throughout Africa, every elephant is important and none should be killed.

Those who decry these officially sanctioned elephant killings include some prominent elephant researchers as well as various animal rights groups, such as the International Fund for Animal Welfare and the Front for Animal Liberation and the Conservation of Nature (FALCON). Some opponents of the killings believe that the elephants should be allowed to reach their own natural population sizes in the parks they inhabit; it is argued that if they are left alone, they will limit themselves. Furthermore, in some people's opinions, elephants are highly intelligent, social creatures that merit more consideration than "bugs," for instance; if there are too many bugs it is emotionally easier to kill a few. Elephants have complex behavior patterns and interactions; they communicate with each other in a manner that some researchers contend borders on language in a human sense. Elephants exhibit altruism, family bonding and structure, and many other characteristics shared with human culture and society. It has even been suggested that if we begin to accept the culling of animals as intelligent as elephants, this may pave the way for the culling of humans if governmental authorities decide that the human population has finally exceeded the capacity of available resources. If in some areas there really are too many elephants, opponents of culling suggest that elephants should be captured alive and moved to parts of Africa where there is a paucity of elephants.

Proponents of controlled elephant cullings point out that such intervention has actually allowed the increase and then stabilization of elephant populations in selected areas. In the 7,700-square-mile Kruger National Park, due to heavy trophy and ivory hunting in the nineteenth century, there were only twelve elephants in 1903. Today, using careful culling to manage the elephants, the population is maintained at a stable 7,500. In South Africa as a whole there were approximately 8,000 elephants in 1981 and 11,000 in 1993 (numbers prior to 1981 are unknown, but they were considerably lower than 8,000). In Botswana there were 28,000 elephants in 1975 and by 1993 this had increased to 65,000. Proponents of culling cite such statistics as evidence that careful elephant management, including culling, can augment elephant populations.

Proponents of elephant culling also contend that elephant overpopulation in a confined area (such as a park or habitat preserve) can not only lead to problems for the elephants, such as disease and starvation, but will have devastating effects on the vegetation and other animals. An adult African elephant typically eats about 220 pounds of vegetation a day. They can push down and destroy trees; huge herds of elephants have the ability to convert woodland to shrubland or grassland, and if there are really too many elephants, the end result may be desert. Such activity on the part of the elephants can lower the natural diversity of other organisms in the area. Various plants will go extinct locally, and with them numerous types of animals, including bird and insect species, will disappear. At Kruger National Forest the authorities explicitly maintain the elephant population at slightly lower than the maximum possible number explicitly to preserve a mixed habitat that will maximize biodiversity in the area.

As for the suggestion that elephants should be captured and relocated, most culling proponents have no problem with this idea if it can be done economically and feasibly. However, it is extremely expensive to capture elephants alive and move them substantial distances. Furthermore, as Africa becomes increasingly populated with humans there are fewer places to move the elephants to. Another alternative to culling might be the introduction of artificial elephant birth control. At present, however, research into elephant contraceptive methods are still in the preliminary stages and it is not expected that such techniques could replace culling or relocation until well into the twenty-first century.

Questions

1. Critically evaluate the argument that since elephants are large, intelligent, magnificent creatures who exhibit many human characteristics, they should be afforded special treatment. Is it somehow "wrong" to cull elephants but not insects or trees?

2. Which is more important: to save the life of an individual elephant, or to protect the overall well-being of an entire population? How does this question bear on the issue of elephant culling?

3. If elephants are not culled in certain parks then the populations of many other species, from other large mammals to plants and insects, may be negatively impacted. What responsibility do humans have to protect these other organisms from elephant overpopulation?

4. The idea of relocating elephants is seen by some as the best alternative to culling elephants, but who should bear the significant economic burden of relocating hundreds or thousands of elephants? What happens when there is no more room to relocate them? In 1994 the International Fund for Animal Welfare paid for 150 elephants to be removed from Kruger National Park and relocated to other parks in South Africa, but hundreds of elephants were still culled. Do you think this money was well spent, or are there more pressing animal welfare issues that it could have been spent on?

Source

Bartlett, Ellen, 1995, "Culling the herd: With elephants in decline, critics decry killings in overcrowded park." *The Boston Globe* (July 24, 1995): pp. 25, 27.

-22-

Should Gulls be Poisoned to Make Way for Other Birds?

The Monomoy Islands are a long, narrow pair of islands that lie off the "elbow" (Chatham area) of Cape Cod, Massachusetts; they are home to the federal Monomoy National Wildlife Refuge. For decades the U.S. Fish and Wildlife Service has attempted to encourage the breeding of various threatened and endangered shorebirds, such as piping plovers and terns, within the refuge at the expense of herring gulls and black-backed gulls. The problem is that the gulls are large and aggressive and out-compete more "desirable" birds. Until 1980 the Fish and Wildlife Service would periodically poison the gulls in the refuge using DRC 1339, a specifically formulated gull poison delivered in bread cubes dropped onto gull nests. DRC 1339 causes kidney failure in gulls, but if used properly affects virtually no other birds. Poisoning of gulls in Monomoy was halted in 1980 after there was a general public outcry against the killing of 2,000 gulls. Many of the poisoned animals flew from Monomoy to the tourist town of Chatham and there expired in the streets--not a pretty picture, and not much help for an economy that relies heavily on tourism.

From 1980 through 1995 the Fish and Wildlife Service attempted to discourage gull breeding on the Monomoy Islands by driving them off using loud noises, such as gun blasts, and dogs. Still, the gulls were not easily discouraged and by early 1996 approximately 25,000 gulls were nesting in the refuge while only 14 plover nests could be found on the islands. In desperation, the Fish and Wildlife Service announced plans in February 1996 to poison 4,000 gulls on South Monomoy Island (the island farthest from Chatham and the rest of the Cape Cod mainland)--an announcement greeted with opposition by some environmentalists, including officials of the Massachusetts Audubon Society. The actual poisoning began in May 1996.

Audubon contended that the plover population on the islands was naturally increasing, rising from 5 nests in 1993 to 14 nests in early 1996, while the gull population had fallen by a third. Furthermore, Audubon officials charged that the Fish and Wildlife Service's actions were primarily politically motivated--by killing gulls and promoting piping plovers and other birds on South Monomoy island, this would take pressure off the mainland beaches to protect plovers. The mainland beaches are heavily used by tourists, and sometimes plovers are run over by off-road vehicles. Indeed, concurrently with the federal moves to kill gulls and increase plover populations on South Monomoy Island, Massachusetts officials were suggesting that state laws might be relaxed so that off-road vehicles could drive closer to plover nests. State officials argued that the plovers had already made a strong comeback on the mainland. In a larger context, Edward Moses, the manager of the Monomoy National Wildlife Refuge, stated that U.S. Interior Secretary Bruce Babbitt promoted the protection of endangered species on federal land so as "to take pressure off non-federally owned property." (quoted in *The Boston Globe*, February 23, 1996, p. 23)

Questions

1. Would you favor the poisoning of the gulls to promote populations of plovers and other birds? Justify your answer.

2. How impressed are you by the Massachusetts Audubon Society's argument that the plover population is naturally increasing and the gull population is natural decreasing? Has the Audubon Society taken into account that the Fish and Wildlife Service has been using other methods to discourage the gulls even when not poisoning them?

3. Do you agree with the concept of protecting endangered species on federal land so as to "take pressure off non-federally owned land"? Explain your reasoning.

Sources

Allen, Scott, 1996, "Agency plans to kill gulls on Monomoy." *The Boston Globe* (February 23, 1996), pp. 19, 23.

Allen, Scott, 1996, "Many island gulls dying on Cape; poisoning continues." *The Boston Globe* (May 21, 1996), p. 25.

-23-

Should Mountain Lion Hunting be Legalized in California?

Prior to 1972 it was legal to hunt mountain lions (also known as cougars or pumas) in California. By the early 1970s, however, the California mountain lion population had dropped to an estimated 2,400 and in 1972 a moratorium on further hunting was imposed. In 1990 a "permanent" ban was placed on all mountain lion hunting in the state.

In recent years, however, the mountain lion population has steadily increased to between 4,000 and 6,000, or possibly even higher. With the increase in mountain lions has come an increase in negative interactions between lions and humans. Ranchers have increasingly complained that mountain lions are killing their livestock, people report being stalked by mountain lions, and in a half-dozen incidents since 1990 people have actually been attacked by mountain lions. In two separate cases, both of which occurred in 1994, female joggers were killed by the lions. In January, 1996, a game warden in Cuyamaca Rancho State Park, forty miles northeast of San Diego, shot and killed a mountain lion on the verge of attacking. In this case, killing the lion was legal since it was considered an act of self-defense. Just a few days previously a confrontation with a mountain lion had been reported by a park visitor in the same area.

As a result of growing concerns over mountain lions, in March, 1996 California state senator Tim Leslie sponsored a ballot initiative, known as Proposition 197, to authorize the State Fish and Game Commission to allow the hunting of mountain lions once again. Supporters of Proposition 197 included many law enforcement officials, hunters, various business persons, and ranchers. Proponents argued that the number of mountain lions could easily sustain limited hunting, and furthermore decreasing the number of lions should reduce the frequency of attacks on human life and property. A number of animal rights activists and environmental organizations opposed it, however. Opponents strongly argued that hunting would do nothing to decrease the number of attacks--in fact, in states where mountain lion hunting is allowed the cats will still occasionally attack humans. More importantly, perhaps, opponents of Proposition 197 noted that even under the hunting ban it was still legal to kill or relocate lions that were considered to be a threat to public safety.

Questions

1. In the abstract, do you approve of hunting big cats, such as mountain lions? If you had a chance to go mountain lion hunting, would you participate? Why or why not?

2. Would you wear a coat made out of mountain lion fur? What do you think of a mountain lion head hanging as a trophy in a living room or den?

3. Should mountain lions be "punished" for attacking humans? If so, how? Should those known specifically to have killed a human be hunted down? Or should it be open season on all mountain lions? Why do you think mountain lions occasionally attack and kill humans? Are they simply following their hunting instincts?

4. If you were voting on California's Proposition 197, which way would you vote? Why?

Postscript

On March 26, 1996, the California electorate voted not to allow the hunting of mountain lions.

Sources

Associated Press, 1996, "Cougars Under Fire on California Ballot." *The Washington Post* (March 18, 1996).

Ayres, B. Drummond, 1996, "Cougars and Lawyers Win in California Ballot Measures." *The New York Times* (March 28, 1996), p. B12.

-24-

Overcrowding in the National Parks--Yosemite National Park as an Example

Yosemite National Park covers nearly 1200 square miles along the western edge of the Sierra Nevada mountain range in central California. It is an area of legendary beauty, containing the Yosemite Valley with its spectacular waterfalls and glacier-sculpted granite peaks and lakes. The area of the present park was inhabited by Native Americans for centuries, and perhaps millennia, but they were driven out by non-Indian "white men" in the 1850s. The naturalist John Muir was a great admirer of Yosemite, even writing a book called *The Yosemite*, and campaigned to establish the area as a National Park--a goal realized in 1890.

The park contains dozens and dozens of species of vertebrates, including fishes, amphibians, reptiles, birds, and mammals. Some of these are rare or endangered. Several hundred black bears live in the park, and sometimes they conflict with human visitors when they try to steal food from campers. Large predators that were thought to frighten visitors, such as wolves and grizzly bears, were purposely eliminated from the park by the Park Service in the late nineteenth and early twentieth centuries, although there is now a general movement toward restoring such top predators to their former habitats throughout the National Park System. Bighorn sheep, which had disappeared from Yosemite by 1914 due to disease, hunting, and habitat destruction outside of the park, were reintroduced to Yosemite in 1986. Among other notable animals in the park are some endangered peregrine falcons. Despite DDT having been banned in the United States since 1972, the Yosemite area is badly polluted with DDT which causes the shells of the falcon's eggs to be very thin and break easily. In order to deal with this problem, the Park Service removes newly laid eggs and replaces them with artificial eggs. The falcon eggs are artificially incubated until they hatch, when the new chicks are returned to the parents to be raised in the "wild." Yosemite National Park also is home to a huge diversity of plant species, including an estimated 1200 flowering species and several dozen species of trees. About 45 of the plant species are considered to be rate or endangered.

Since its founding as a National Park, Yosemite has proven to be quite popular. Easily accessible by car from such major population centers as Los Angeles and San Francisco, even in the first decades of the twentieth century Yosemite was inundated by hordes of visitors. The early Park Service did everything they could to encourage visitors, for this was the purpose of a National Park. The Park Service eliminated animals, like the wolves, that would scare visitors and encouraged other animals, like deer, to flourish. They built roads, paths, lodges, restaurants, and other amenities so that the park could be enjoyed in peace and comfort. They tried to put out naturally occurring forest fires whenever possible. The wild and untamed aspects of the park that John Muir and other early naturalists had so much appreciated were mitigated, and visitors flocked to Yosemite in ever increasing numbers.

With the growing numbers of tourists all sorts of problems started to develop in Yosemite. Litter, erosion along trails, reduction of the natural wildlife, pollution, noise, overcrowding, and commercialization were all becoming evident before 1920 and only worsened as the century progressed. People came to Yosemite to enjoy the quiet solitude, the pristine beauty, of nature--yet the increasing number of visitors was eliminating those very characteristics of the park. To serve visitors various restaurants, service stations, hotels and lodges, swimming pools, tennis courts, kennels, liquor stores, and even a miniature golf course and shopping center were located inside the park. By the 1970s through 1990s millions of visitors were entering the park every year--about 3.5 million in the early 1990s. Such huge crowds could only lead to problems. More and more cases of vandalism, theft, speeding, drunk driving, and other violations were reported. Park rangers typically had to deal with ten to twenty thousand incidents per year in the 1980s and 1990s.

Over the July 4th, 1970 weekend over 76,000 people crowded into Yosemite Park at one time. Several thousands of these, belonging to the younger generation, occupied a meadow and played loud rock music, used drugs, danced, and so forth. Park Service officials tried to politely break up the crowd, but their initial attempts were unsuccessful. Park Rangers ultimately had to break up the crowd using clubs and Mace, arresting 186 people in the process. More and more the de facto role of the Park Rangers was to be police officers rather than guides and naturalists.

In the 1970s and 1980s the problems only seemed to get worse. Memorial Day, 1985, saw major traffic gridlock in Yosemite, and in August 1990 forest fires temporarily trapped an estimated 10,000 tourists in Yosemite Valley. Fortunately, they were evacuated after about a day, but the fire forced the park to be closed down for ten day and an estimated 25,000 acres burned.

In order to deal with the problems of overcrowding and degradation of the park, the Park Service is in the process of changing its approach. They are working on ways to control, and limit, the number of visitors to the park--both day visitors and overnight visitors. Commercial exploitation of the park is being curtailed; for example, certain types of non-essential concessions (for example, some gift shops and excess sandwich stands) are being eliminated. As far as managing the natural aspects of the park, the trend is now to reintroduce all species that originally lived in the park, including top predators, while eliminating exotic species (especially plants) that have invaded the park. Areas stripped of vegetation are being replanted, and grazing of domestic animals, such as burros, horses, and mules, is more strictly regulated. Also, the Park Service no longer suppresses all natural forest fires--some are allowed to burn. In fact, controlled fires have been purposefully set to burn off dead wood in some areas that accumulated as a result of the decades-old policy of putting out all forest fires.

Questions

1. Yosemite Park is no longer a truly pristine, wild, natural habitat; in fact, it is carefully managed by the Park Service. The Park Service has to control prey species, such as elk, since populations of their natural predators--for example, wolves--have been either eliminated or seriously curtailed. The Park Service traditionally controlled forest fires in the park, they monitor peregrine falcon eggs, and so forth. In your opinion, is such management necessary or desirable, or should the park simply be allowed to follow its own "natural development" with minimal human intervention?

2. What do you think the purpose of a National Park should be? To preserve pristine nature for those that love nature? To preserve nature for its own sake? To serve the recreational needs of a public who may prefer their "nature experience" tempered by good roads, easily walked trails, nice hotels, gift shopping, fancy restaurants, and so forth?

3. Since it is owned by the United States citizens, should there be restrictions placed on the number of people allowed into a park? Is it elitist, or even discriminatory, to suggest that parts of a National Park should remain relatively wild and therefore accessible only to the relatively young and able-bodied? Or should the majority of a National Park be made easily accessible to all, young and old, healthy and handicapped?

4. Once a concessionaire, such as a restaurant operator, has invested a significant amount of time and money into a concession within a National Park, should he or she be forced to move just because environmentalists and conservationists feel the concession detracts from the "natural setting"? Does it matter if business is brisk, therefore indicating a desire among the general public for the services?

Sources

Franck, Irene, and David Brownstone, 1992, *The Green Encyclopedia*. New York: Prentice Hall General Reference.

Kaufman, Donald G., and Karla Armbruster, 1993, *Finding Our Niche: The Human Role in Healing the Earth*. New York: HarperCollins College Publishers.

-25-

An Excess of Deer

For many years people have worried that wildlife is being decimated by humans, and while this is a legitimate concern in numerous areas, in some regions certain types of wildlife have rebounded with a vengeance to plague humans. Such is the case with deer in Northern Virginia and Maryland around the area of Washington, D.C. In Maryland the deer population has exploded in the last seventy years--in 1931 a mere thirty-two deer were taken during the hunting season, compared to 51,000 in 1993. The lack of deer taken in 1931 was not due to a dearth of deer hunters. In 1995 the estimated deer population was over 200,000.

While some people might think it cute to have wild deer frolicking in their yards, for many communities in Montgomery County, Maryland, deer have become a major nuisance. They have invaded this suburban area just northwest of the nation's capital, destroying flowers, shrubs, and trees. They have become the bane of the serious and amateur gardener alike. The deer seem to favor azaleas, rhododendrons, roses, daylilies and true lilies, hostas, columbine, yews, hemlocks, arbovitae, Japanese maple trees, and winged euonymous, to name but a few of their favorites; however, when hungry enough (especially during winter months), deer will devour almost any garden vegetation. The deer are amazingly bold--in some cases they will feed on plants that are growing right against the side of a house. To make matters worse, during rutting season (approximately October to January), young male deer polish their antlers against trees. They prefer smooth, young trees and often seriously damage or destroy the trees in the process.

Deer cause damage not only to lawns and suburban gardens, but also to agricultural crops and garden nurseries. Indeed, in Maryland and Virginia they have caused hundreds of thousands, and perhaps millions, of dollars worth of damage--and probably an equal amount has been spent on dealing with the deer problem. Then there are other types of problems associated with the deer. Deer are carriers of the tick that spreads the potentially deadly Lyme disease to humans, and deer are involved in many traffic accidents. The deer, who become accustomed to people and traffic, wander onto major roads and are hit by unsuspecting motorists. In 1993 946 deer were killed in road accidents in Montgomery County, in some cases causing major damage to vehicles, and injuries and even death to human occupants.

The deer problem can be attributed to a number of factors. Humans have eliminated the deer's natural predators, such as bobcats and coyotes. Out of compassion for the deer, hunting has been severely restricted or even banned in many areas. Deer habitat has been increasingly encroached upon by human development such that the deer have nowhere to go but into human neighborhoods. Furthermore, during development mature trees and forests are routinely destroyed--it is the mature forests that can best sustain populations of deer. A small herd of deer can feed on the acorns of a large, old oak without damaging the tree. When the deer are forced to feed on newly planted shrubs and flowers they often destroy them completely.

Various strategies have been devised to try to control the damage done by the deer. Gardeners invest in plants that deer tend not to favor, although they may eat them anyway. Gardens and yards are fenced in. Some people favor electrified fences, but these can be dangerous to humans, especially young children. Dogs, chemical repellents, and mechanical devices (such as strobe lights, high-pitched sound emitters, and devices that automatically spray deer with water) are used by some, but all have disadvantages. Such "solutions" are often not permanent as the deer become accustomed to them. In one case a product manufactured from coyote urine scared away the deer for a time--until the deer figured out that there weren't really any coyotes in the area. Also, any such deer repellent chemicals or devices are often a nuisance, or even a danger, for humans. Authorities have had limited success with trapping and relocating nuisance deer--more always seem to replace those removed. Limited attempts at deer contraception have yet to yield significant results.

There has been growing support among those afflicted by deer problems for increased hunting as the best way to control the population. But such solutions have also been vocally opposed. For instance, a group called the Fund for Animals has fought against deer hunting, suggesting instead that gardeners use more fences and/or plant extra vegetables for the deer. The latter solution, critics argue, will only attract more deer and help maintain their high numbers.

Questions

1. In your opinion, which do you think is the worst problem: that deer destroy gardens, that deer spread Lyme disease, or that deer cause traffic accidents?

2. What kinds of solutions to the suburban deer problem in Montgomery County, Maryland, could you support? Would you advocate increased deer hunting, for instance? Is it realistic to hunt deer in suburban neighborhoods? In some cases hunters can use crossbows in certain areas, and during certain seasons, to take deer even if the use of firearms is illegal. Do you think crossbow hunting is preferable to hunting with rifles or shotguns?

3. Are the deer to be blamed for damaging human gardens and otherwise being a nuisance? What are the causes of the deer population explosion? Should the deer suffer for a situation that they did not intentionally create?

Source

Higgins, Adrian, 1996, "When Deer Invade the Garden." *The Washington Post* (February 8, 1996, "Washington Home" Section), pp. 8-11.

-26-

Saving the Galapagos

The islands of the Galapagos (off the coast of Ecuador, and belonging to that country) fascinated Charles Darwin. The collection of over a hundred islands contain a unique fauna and flora that helped Darwin to develop and refine his theory of evolution by means of natural selection. The giant tortoises of the islands, bats, iguanas, fur seals, penguins, finches, mockingbirds, cacti, acacias, and numerous other indigenous species of the Galapagos have been viewed as classic case studies illustrating evolutionary principles ever since Darwin's time. Active research in this natural laboratory continues to this day. But Darwin was not the first to admire the islands: they were named Las Encantadas (the Enchanted Isles) in 1535, and they have attracted tourists for their scenic beauty ever since. The uniqueness, scientific importance, and beauty of the islands has been formally recognized in recent times: 97% of the land area of the islands is designated as a national park, the ocean in and around the island archipelago is designated a Marine Resources Reserve, the inland waters of the area are an International Whale Sanctuary, and the islands are a United Nations Educational, Scientific, and Cultural Organization World Heritage Site and also a U.N. Man and the Biosphere Reserve.

For centuries after the islands were discovered their human population remained extremely sparse. As recently as 1970 only 2,000 people lived on the islands year-round, and they entertained about a thousand tourists and guests each year. Now the largest town alone has a population of about 8,000 and there are some 15,000 permanent residents spread over the islands, and the population is growing by some 8% a year. Additionally, 50,000 visitors make their way to the Galapagos annually. Many of these visitors are ecotourists or scientists who are generally careful to help preserve the natural environment, but increasingly more of the visitors are simple recreationists who tend to have less interest in the unique species and habitats of the Galapagos. Furthermore, some of the tour operators and ship crews servicing the tourist trade pay little heed to the natural environment--for example, they may dump garbage and sewage into the ocean and leave trash on the islands. The increase in tourists, and the increase in people to work in the tourist industry has severely strained the freshwater supply, sewage systems, and general infrastructure of the Galapagos. Additionally, the traditional social structure of the long-time residents of the islands is being strained by the influx of newcomers, many of whom are there only to make money. With the newcomers, it is alleged, has come an influx of drugs, prostitution, and other social problems.

These new people have brought with them numerous alien species that are now threatening the fragile Galapagos ecosystems. Goats, burros, dogs, cats, cows, rats, fire ants, and wasps, all imported from the mainland, are destroying the native flora and fauna. An estimated 50,000 to 75,000 goats now live on Isabela Island, the largest island of the Galapagos, and feed on the native vegetation on which the tortoises are dependent. To make matters worse, burros trample the tortoise's nests, and dogs feed on the native iguanas. Across the Galapagos islands there are an estimated 300 species of alien plants now competing with the native vegetation--and of these 300 species, approximately a third have been introduced since 1985.

Not only do goats, dogs, and other animals prey on the native species, but so do humans. Fishermen go after lobsters, sharks, sea cucumbers (a delicacy in Asia), pipefish, seahorses (considered an aphrodisiac), sea urchins, and sea lions in the waters around the islands. While some of this fishing and hunting is legal, much of it is illegal. To make matters worse, fisherman have been illegally camping in the national parkland areas, have cut and burned mangroves in order to dry and cure the sea cucumbers they catch, and have even made meals of the giant tortoises.

In response to such abuses, in 1994-1995 the Ecuadorian government closed the fishing season a month early, which set off a string of protests on the islands. Local fishermen and other natives dependent on the industry blocked the entrance to the Charles Darwin Research Station in Puerto Ayora on Santa Cruz Island, holding the workers and scientists inside captive for four days. Later in the same season locals held the station for two weeks and also occupied an airport on the island San Cristobal. Many local residents of the Galapagos are demanding more autonomy from the mainland government. Some observers fear that there could be an actual rebellion on the islands, and if this occurs the local habitat could face even further danger. Already local leaders have threatened to take tourists hostage and burn parts of the parks and sanctuaries if they are not heeded. Currently negotiations between the islanders and mainland government officials are proceeding; it is difficult to predict what the outcome will ultimately be. If the natives gain significant political autonomy, as they desire, they may use it to loosen regulations currently in place to protect the unique biota of the Galapagos islands. Such a result might be a victory for the local people, but a disaster for the native non-human residents.

Questions

1. Is it possible to save every set of islands in a pristine state? Would this even be a desirable goal? Is there any particular reason that the Galapagos islands should be singled out? Are they inherently more important than other parts of the globe? Or do they simply hold a special place in the hearts of biologists due to Darwin's exploration of the area?

2. Most of the Galapagos has been designated as park, reserve, or sanctuary--are such

designations adequate in and of themselves, or must the area be continually monitored? What do you think might become of these areas if the Galapagos gain further autonomy from the Ecuadorian mainland government?

3. Discuss the current social unrest found on the Galapagos islands. What is the cause of this unrest? How might it be resolved? What effect is it having on the native flora and fauna?

4. Do you think there is a problem to be solved on the Galapagos? Or should the islands simply be allowed to change without impediment as more people inhabit them and visit them? Justify your answers to these questions. If you think there is a problem on the Galapagos, what solution would you suggest? Should the Ecuadorian government crush the local autonomy movement and institute strict controls to preserve the native flora and fauna? Should immigration to the islands be limited? Should the annual number of visitors be limited?

Sources

Bowler, Peter J., 1984, *Evolution: The History of an Idea.* Berkeley: University of California.

Lemonick, Michael D., 1995, "Can the Galapagos Survive?" *Time* (October 30, 1995), pp. 80-82.

-27-

Smallpox: Should It Be Eradicated Completely?

Smallpox, caused by the variola virus, afflicted humanity for centuries. In Europe and Asia during the fifteenth through eighteenth centuries towns were sometimes decimated by periodic smallpox epidemics. With the discovery of the New World Europeans introduced the disease to the Americas with devastating results--millions of Native Americans died. Victims of the disease suffered from fevers, erupting lesions, and pus-filled blisters (most commonly found on the face, forearms, and wrists). Depending on the area and the severity of the smallpox strain, 60% to better than 95% of those who came down with the disease might survive. But those lucky enough to survive were usually scared for life with deep pits (the "pox"), and in some cases also went blind. In terms of deaths caused to humans, smallpox may have been the worst disease in history.

In 1796 Edward Jenner, an English physician, introduced a successful vaccination for smallpox derived from the related but much milder disease cowpox. The spread of smallpox vaccinations during the nineteenth and early twentieth century helped limit the damage caused by the disease. However, in the 1960s smallpox still infected an estimated 2.5 million people a year (mostly in the developing world). Furthermore, with the increasing popularity of ship and especially airline traffic around the world there was serious concern that a single person might contract the disease and transmit it to an area where smallpox had been eradicated, thus causing a major epidemic. In order to avoid such as disaster, in 1967 the World Health Organization began a concerted effort to totally eliminate smallpox around the world. Ten years later the job was essentially finished; the last case of "wild" smallpox was reported in late 1977. The only known reservoirs of variola virus were safely locked away in medical laboratories.

This might have been the end of the story, except . . . in 1978 the virus somehow escaped from a medical laboratory at the University of Birmingham, Great Britain. It infected two people, a medical photographer (who died of the disease) and a laboratory supervisor who committed suicide. As a result of this incident, a number of labs destroyed their last remaining stores of variola virus. As of 1983 the only stocks of variola virus were under tight security at the Centers for Disease Control in Atlanta, Georgia, and at the Research Institute for Viral Preparations in Moscow. The virus was studied intensely, and maps of the entire variola genome were prepared. With this research completed, the World

Health Organization proposed that the last reservoirs of variola virus be destroyed on December 31, 1993. Such a move would be the first absolutely premeditated, controlled, deliberate total extinction of a "life form" (actually, in many respects viruses are not "alive," at least in the traditional sense--but they do share many biochemical traits with undoubtedly living organisms).

The proposed extinction of the variola virus was vehemently debated at an international virology conference held in Glasgow, Scotland, in August 1993. Those arguing against such deliberate extinction felt that once the virus is gone, it is gone forever and one can never predict what knowledge might be gained from it in the future, or even what uses it might have. Through tighter security arrangements, another incident like that which occurred in 1978 should be avoidable. Those arguing for extinction felt that the risks of maintaining the variola virus were too great. Now that it is eliminated in the wild, humans no longer need to be vaccinated for smallpox and have no general immunity to the disease. If the virus escaped, it could cause major devastation around the world.

Questions

1. What do you think is the strongest argument in favor of preserving the smallpox-causing variola virus? Given that we have only a limited understanding of many viruses and their potential importance, should samples of the virus be preserved for posterity? What are the implications of destroying something that we do not fully understand?

2. What do you think is the strongest argument against preserving the virus? What if the virus somehow fell into the wrong hands, such as a terrorist organization? What if the virus simply escaped by accident, as it did in 1978?

3. Some people argue, primarily from a moral or spiritual perspective, that humans have no "right" to totally destroy any species. Can and should the argument be applied to a disease like smallpox? Why or why not? How might your religious background and social upbringing affect your answers to these questions?

4. Ultimately, would you support the controlled extinction of the variola virus or not? Justify your position.

Postscript

The destruction of the final remaining stocks of variola virus has been repeatedly postponed, but currently they are scheduled to be destroyed in June 1999. With their destruction, it is believed that smallpox will have been completely eradicated. However, with approximately six billion people on the planet, it is always possible that a small pocket of smallpox disease remains among the inhabitants of some remote area.

Sources

Anonymous (Reuters), 1996, "WHO sets June 1999 date to destroy smallpox stores." *The Boston Globe* (May 25, 1996), p. 80.

Davis, D. J., 1965, "Smallpox." *Encyclopedia Americana, International Edition* (New York: Americana Corporation.) 25:107-108.

Joklik, W. K., *et al.*, 1993, "Why the Smallpox Virus Stocks Should Not Be Destroyed." *Science* 262:1225-1226.

Mahy, B., *et al.*, 1993, "The Remaining Stocks of Smallpox Virus Should Be Destroyed." *Science* 262:1223-1224.

Mann, Charles C., and Mark L. Plummer, 1995, *Noah's Choice: The Future of Endangered Species.* New York: Alfred A. Knopf.

-28-

Saving Wood

One of the reasons that the forests are disappearing so quickly (only 1.5 billion hectares of undisturbed primary forest remain out of an original 6.2 billion hectares, and 17 million hectares of tropical forest alone continue to be lost each year) is that demand for wood is higher than it has ever been, and the demand continues to grow. Unfortunately, throughout the world forests are cut down unsustainably--that is, the forests are cut faster each year than they can grow and replenish themselves. To give just a few examples, in China forest cutting exceeds regrowth by 100 million cubic meters of wood each year. In India an estimated seven times as much wood is taken from the forests each year as can regrow; thus the Indian forests continue to disappear at an alarming pace. Such unsustainable cutting of forests is not limited to developing and Third World countries. In 1989 logging in British Columbia (Canada) was 30% higher than the sustainable yield, and on the West Coast of the United States during the 1980s harvests were about 25% over sustainable yields on commercially-owned lands and about 61% over sustainable yields on federally-owned lands. In Washington, Oregon, Alaska, and elsewhere federally owned forests are being plundered for the benefit of a few lumber corporations and pulp mills, including Japanese-owned corporations. To add insult to injury, much of this destruction is subsidized by American taxpayers. For example, in Alaska's Tongass forest the United States Forest Service lost more than $350 million during the decade 1980-1990 from sales of timber (more was invested in roads, construction, and so forth, to open the forest to lumbering companies and supply pulp mills than was made from the sales of timber), and most of the cheap timber headed to Japan.

Some of the wood taken from the world's forests is used as fuelwood, and another large portion is harvested for industrial purposes. On a global scale, of the timber that is taken commercially, over half goes toward the building industry. Half of all industrial wood is sawn into lumber, and over an eighth is made into plywood, chipboard, and similar products. The second largest industrial use of wood is to turn it into pulp to manufacture paper and related products. The remaining wood is used for furniture, carvings, handles for tools, and thousands of other uses.

Much of the commercially harvested wood is simply wasted. During initial logging operations enormous amounts of wood are sometimes lost, including during transport. In many areas logs are floated down rivers, and not all of the logs make it to their destination. In the former Soviet Union it has been reported that numerous logs are lost during such transportation--perhaps up to 3 million cubic meters of wood per year.

In sawmill operations any leftover wood chips and sawdust are generally used in pulp mills or burned as an energy source. In this regard little wood is actually wasted, but it is still important to maximize the amount of solid wood product that can be produced from freshly harvested trees. Any losses in sawdust used for pulp or fuel can be made up by recycling paper and other wood-based products. During cutting and manufacturing waste is also rampant using traditional methods. In contrast, new, thinner saw blades used to cut logs into boards save wood, computers can scan each individual log and determine the best way to cut it to maximize the usable wood, lathes can peel logs for manufacture into plywood more efficiently, and so on. The United States Forest Service has estimated that the increasing price and demand for lumber will, by the year 2040, reduce the average amount of raw wood used to produce a final piece of lumber or sheet of plywood by at least one third.

Perhaps the best way to save wood is by changing our consumption habits. It has been argued that in the United States most new houses, built primarily of lumber and wood products, are "overbuilt." Beams could be moved apart to reduce lumber requirements, slightly thinner wood members could be used in many instances, and so forth, without noticeably affecting quality or safety. The U. S. Forest Service has estimated that in this manner 10% of the lumber used to build the typical new American house could be saved. In addition, hundreds of kilograms of wood are typically lost as waste (such as the ends of boards being sawn off and discarded) during construction of a house. In Japan the use of plywood manufactured from high-quality Southeast Asian tropical hardwoods is notorious. A substantial amount of this plywood (up to one third) is used to make panels for the molding of concrete--after only a few uses of the panels, the plywood is discarded. Japan also uses an estimated 20 billion disposable wooden chopsticks a year. Instead of disposable concrete molding panels and chopsticks, better-quality reusable items could be utilized.

Disposability is one of the hallmarks of western, industrialized societies--and nowhere is this documented better than in our wasteful consumption of non-durable paper products. The use of paper and paperboard products--everything from toilet paper, stationery, advertising brochures, newsprint, cardboard boxes, paper plates and napkins, disposable baby diapers, and thousands of other products--has increased dramatically in the twentieth century. Much of the increase is due to modern packaging of foods and other products, and to advertising in all of its printed, paper-consuming guises. In the United States the per capita annual consumption of paper and paper products was 317 kg in the late 1980s, and half of this paper was in the form of packaging. The vast majority of paper is made from wood pulp (rag paper is a rare commodity used only for special purposes). Some of the pulp comes from sawmill residues, and some of the paper is recycled from used paper, but still large tracts of forest around the world are logged simply to make pulp for paper products, including forests in Canada, Southeast Asia, and the Tongass National Forest in Alaska.

Perhaps the easiest and most immediate way to cut down on wood use, thus saving trees and forests, is to reduce paper consumption and increase the use of recycled paper where paper products are necessary. This is easier said than done, however. The modern information age, the widespread use of computers, and the invention of inexpensive and reliable photocopying machines have only served to increase paper consumption. Businesses send out huge mailings--45 billion pieces of bulk mail were handled by the U. S. Postal Service in 1986--and modern electronics have not proven a substitute for paper. In some cases the need for paper has even increased further. Virtually everything that goes into computers, electronic financial transactions, and so forth, comes out eventually on paper. It seems that people generally do not trust electronically stored records, and they prefer to read from a piece of paper rather than a computer screen. Credit cards generate paper and mailings; bills and statements must be sent. U.S. banks process tens of billions of canceled checks annually. There are more and larger newspapers, magazines, and periodicals in circulation than ever before. It seems to be more difficult to wean the public away from excessive paper packaging than might have been anticipated. To top it off, the market for recycled paper experiences tremendous swings; in the late 1980s and early 1990s large quantities of paper ready for recycling could not find buyers--the technology and economics of paper manufacturing made it more viable for many companies to produce paper from original wood pulp--so the forests continued to be destroyed. Fortunately, however, this situation has now been reversed and paper for recycling is a hot commodity.

Still, there are simple things that everyone can do which will make a big impact. Many constructive ways to reduce paper consumption are nothing more than common sense. Use double-sided photocopying, share newspapers and magazines with your neighbors (why should each individual buy a paper every day, only to barely glance at it and then discard it?), stop delivery of unwanted bulk mail (write to the company that is sending it, or to a service that will take you off unwanted mailing lists), reuse envelopes and paper with a blank side, avoid disposable paper picnic goods, buy only recycled paper (this is good in and of itself, and will also further encourage the market for such products), refuse to buy over-packaged goods, reuse paper grocery bags (or better yet, use permanent cloth bags), and finally don't be afraid to think of other creative ways to save paper and thus save trees. Saving paper not only saves trees, but saves fossil fuels (it has been estimated that it takes about 1,500 pounds of petroleum to produce a ton of paper), cuts down on air pollution and the emission of greenhouse gases, and cuts down on solid municipal wastes (paper products make up about a third of all solid municipal wastes in the United States).

Not only is overharvesting of forests destroying them, but so is pollution--so anything we can do to cut down on pollution will also help to save forests. In Europe and elsewhere around the world air pollution and acid precipitation are causing measurable damage to forests. It is estimated that in Europe a minimum of $30 billion a year in damage to forests is caused by pollution--trees die or do not grow as well. If the greenhouse effect develops as most scientists predict, this will have a devastating impact on the forests of the world, especially temperate forests which will not be able to tolerate the rising temperatures.

Questions

1. How serious do you think the forest situation really is? Do we really need to worry

about running out of wood? Are there reasons to preserve wood and forests other than the fear that our future wood resources will be depleted?

2. What do you think should be done to save wood and protect the forests of the Earth?

3. Cutting corners in home construction as a way to save wood has been criticized as a dangerous practice--houses may result that are not structurally sound and could collapse. Do you think there is any truth to such a suggestion?

4. Wood is totally natural, easily recycled to a certain extent (such as turning old wood products into paper), fully biodegradable, and a renewable resource--so what is wrong with using wood products? Shouldn't we be encouraging the use of wood over non-renewable materials, such as plastics or metals?

Sources

Brown, Lester R., Christopher Flavin, and Hal Kane, *Vital Signs 1992: The Trends that are Shaping Our Future*. New York: W. W. Norton.

Meadows, D. H., D. L. Meadows, and J. Randers, 1992, *Beyond the Limits: Confronting Global Collapse, Envisioning a Sustainable Future*. Post Mills, Vermont: Chelsea Green.

-29-

Spotted Owls and Old-Growth Forests in the Pacific Northwest

The northern spotted owl has come to symbolize, in some people's minds, the struggle between radical or elitist environmentalists who want to preserve nature and the honest working people who need to make a living and cannot afford to bear the burden of environmentalists' whims.

The northern spotted owl was officially classified as a threatened species by the U.S. Fish and Wildlife Service in 1990, and in May 1991 the Fish and Wildlife Service advanced a proposal to set aside 11.6 million acres of old-growth forest (presumably the northern spotted owls' primary habitat) in the Pacific Northwest in order to protect the owl. Those in the timber industry vehemently objected, and three months later the Fish and Wildlife Service reduced the proposed habitat to 8.2 million acres. Those in the timber industry continued to object as this would put off-limits prime timber stands. Top government officials, industry leaders, environmentalists, and others continued to argue over the relative merits of protecting the owls and the old-growth forests, or the timber industry and the jobs it provides in the area. Sentiments were sometimes expressed on bumper stickers: "Save a Logger--Eat a Spotted Owl" or "Loggers, too, are an Endangered Species." For many people the argument boiled down to "jobs versus the environment." The whole issue became so heated, and important politically, that shortly after being elected President Bill Clinton called a "Timber Summit" to discuss the issues (held in Portland, Oregon, in April 1993).

At the summit both environmentalists and timber company representatives presented their positions; a few months later the Clinton administration suggested a compromise plan. As might have been expected, the compromise pleased virtually no one. The compromise suggested setting up a reserve system to protect many old-growth forest stands, eliminating certain tax loopholes that benefit the forest industry, and providing over a billion dollars in economic assistance to communities that depend on the timber industry (to help retrain and otherwise assist loggers as they lost their jobs). Still, the proposed level of cutting in the national forests was considered to be too high--in fact, unsustainable--by many environmentalists. The timber industry disagreed, arguing that even more cutting than proposed in the compromise could be safely permitted. The controversy remains.

Although sometimes summarized in terms of "owls versus loggers," in many ways the controversy has very little to do with owls (or even loggers) per se. The real argument is over whether, and how much, of the old-growth forest should be lumbered. The northern spotted owl, through its status as a threatened species, served as a vehicle by which environmentalists could attempt to protect the old-growth forests. Simplistically, the argument boiled down to the following: The northern spotted owl as a threatened animal under the Endangered Species Act of 1973 is protected from being hunted, killed, injured, and so forth, plus federal agencies are not allowed to carry out or fund any activity that will endanger the species's habitat. The preferred or required habitat of the northern spotted owl is the old-growth forests of the Pacific Northwest. These same forests, many of which are federal lands managed by the U. S. Forest Service, are prime areas for logging by the timber industry (the Forest Service historically has promoted logging). By demonstrating that these particular old-growth forests are the habitat of the northern spotted owl lumbering can be banned and the forests preserved. Thus the owl is simply a means to the ultimate end of limiting lumbering in old-growth forests.

There have been serious criticisms of such a strategy, especially from within the environmental movement itself. The old-growth forests of the Pacific Northwest are areas containing ancient, giant trees (some over 2000 years old, standing some 300 feet tall and ten feet in diameter) and an associated unique ecosystem. Commercially, the wood in these forests is extremely valuable and, depending on one's method of estimation and definition of old-growth forest, 70 to 95% of the old-growth forest has been logged over the last 150 years. What remains is only a small remnant of the former forest. If logging is maintained at recent rates (those of the late 1980s and early 1990s), all of this forest could be gone only a few decades into the twenty-first century. Protecting these forests may be a worthy goal, but doing it by means of the northern spotted owl may not be the best way.

The northern spotted owl is considered threatened primarily due to its rarity; it has been estimated that perhaps as few as 2,000 breeding pairs exist in the world today (in 1986 the Audubon Society estimated that a minimum of 1,500 pairs would be needed to preserve the owl from extinction). The northern spotted owl is not even a distinct species, but a subspecies of the spotted owl (*Strix occidentalis*); other subspecies include the California spotted owl and the Mexican spotted owl. The California spotted owls and northern spotted owls are particularly close breeds; in fact, some researchers even lump them together as "Pacific Coast spotted owls." If the northern and California spotted owls are for all practical purposes indistinguishable, this argues against the threatened status of the northern spotted owl (since the combined population of all "Pacific Coast owls" may number well over 10,000 breeding pairs). To hang the fate of the Pacific Northwest old-growth forests on the questionable taxonomic distinctiveness of the northern spotted owl may not be advisable.

Even if we accept the distinctiveness of the northern spotted owl, some researchers have suggested that the population numbers for the species have been grossly underestimated. Concentrating mostly on old-growth forests of Oregon and Washington, estimates for the northern spotted owl population have been placed by some authorities in the range of 3,000-4,000 pairs; other researchers focussing on northern California have suggested that there may be 4,000 to 8,000 northern spotted owl pairs in California alone! If this is true, we could clear-cut all the old-growth forest in Washington and Oregon and still have plenty of northern spotted owls in a natural habitat.

The issue of the northern spotted owl's required habitat is also the subject of controversy. Traditional environmental dogma would have it that each pair of owls typically requires a 2,000 to 3,000 acre home range in which to live and reproduce, and this home range needs to be in the old-growth forests. But some owl researchers disagree, pointing to apparently well-established and stable northern spotted owl populations in younger, sometimes forestry-managed forests. If the northern spotted owls do not really require old-growth forests, then the argument for preserving old-growth forests to benefit the owls quickly disappears.

Proponents of logging in the old-growth forests of the Pacific Northwest can attack the defense of the northern spotted owl on such grounds as those discussed above. With this in mind, advocates for the preservation of the old-growth forests may be better off cogently arguing for the forests per se, not the preservation of a particular species that happens to inhabit the forests. However, the legal apparatus to implement such a strategy is generally lacking, which is why the Endangered Species Act has been used as a mechanism in attempts to preserve habitats.

Some observers have pointed out that, due to the activities of certain environmentalists, antilogging sentiments can be raised to such a point among the public that commercial logging even in young forests artificially planted for the express purpose of producing lumber is viewed as villainous. In fact, logging can be carried out sustainably if done properly. This is perhaps most easily accomplished in young forests where as trees are removed new ones are planted, but it can be accomplished even in old, established forests. Such sustainable forestry is best for the environment, the economy, the lumber companies, and the lumber workers. Clear-cutting old growth forest may provide good jobs and huge profits in the short term, but in the long run the forests will be depleted, the jobs will disappear, the profits will end, and the environment will suffer. It is not a matter of "jobs versus the environment" but rather "jobs and sustainability." Owls, forests, and lumber workers can all co-exist together indefinitely. A classic example of such a successful company was the Pacific Lumber Company (founded in 1869) which, from the 1930s to 1985, thrived on a policy of perpetual sustained yield. The company owned its own lands and selectively marked and cut only certain mature trees so as to allow younger trees to mature and replace those cut. As necessary, reseeding was carried out. This approach meant that the Pacific Lumber Company could have continued to produce wood indefinitely. The employees were well treated, with good wages, company-subsidized housing available, room for promotion, good accident and health insurance benefits, a generous pension plan, and high job security. All this ended in 1986 not due to environmentalists or owls getting in the way of logging, but due to the greed of a hostile takeover. Pacific Lumber was purchased with money raised from junk bonds. In order to pay their debts, the new owners raided the employee pension fund and also began clear-cutting Pacific Lumber old-growth forest holdings as a way to raise cash as quickly as possible.

Some defenders of the lumber industry maintain that even if local abuses have occurred and continue (such as completely decimating old-growth forests in order to turn a fast profit), overall the forests of the United States are in very good shape as compared to the recent past. As evidence, they cite statistics such as that New Hampshire was approximately 50% forested in 1850 (in the east, heavy deforestation occurred in the eighteenth and early nineteenth centuries) but is about 86% forested today. Likewise, Vermont was perhaps 35% forested in the late nineteenth century and is 76% forested today, while Massachusetts was about 35% forested in 1850 and is 59% forested today. In the United States as a whole (the same holds true for Western Europe), it is true that more area is covered by forest today than at the turn of the century (about 1900). What these critics of old-growth forest preservation fail to note, however, is that this modern forest cover is second- and third-growth forest. The scrubby forest that has regrown on abandoned farms in New England, for instance, is in no way comparable to the magnificent old-growth forests of the Pacific Northwest.

Questions

1. Do you favor extensive old-growth forest preservation? Or are you of the opinion that only a very limited area, if any, of old-growth forest needs to be preserved (following the adage, "if you've seen one redwood, you've seen them all")?

2. Why was it convenient for some participants in the controversy over the logging of old-growth forest in the Pacific Northwest to latch onto the northern spotted owl? How did this serve to polarize the clash? Whose interests do you think were best served by making the owl a focus of controversy?

3. For a person concerned about preserving old-growth forest, do you think the "owl strategy" is a good approach? Why or why not? If you do not support it, what approach would you recommend?

Sources

Easterbrook, Gregg, 1995, *A Moment on the Earth: The Coming Age of Environmental Optimism*. New York: Viking.

Jefferson, Jon, 1995, "Timmmberr! How Two Lawyers and a Spotted Owl Took a Cut Out of the Logging Industry." In *Taking Sides: Clashing Views on Controversial Environmental Issues (sixth edition)* (edited by Theodore D. Goldfarb), pp. 196-201. Guilford, Connecticut: Dushkin Publishing Group.

Newton, Lisa H., and Catherine K. Dillingham, 1994, *Watersheds: Classic Cases in Environmental Ethics*. Belmont, California: Wadsworth Publishing Company.

Wood, Gene W., 1995, "Owl Conservation Strategy Flawed." In *Taking Sides: Clashing Views on Controversial Environmental Issues (sixth edition)* (edited by Theodore D. Goldfarb), pp. 202-205. Guilford, Connecticut: Dushkin Publishing Group.

Section 3

Problems of Environmental Degradation

-30-

The Rise and Fall of DDT

The pesticide DDT (dichlorodiphenyltrichloroethane, a type of organochlorine) was first synthesized in 1873, but it was not until 1939 that its insecticidal properties were discovered. Seventy-five years ago the world faced many of the same problems we face today--a rapidly increasing world population which means more mouths to feed, problems with pests that diminish the potential of agricultural production, and diseases that are spread by insect vectors such as malaria. In order to help address such problems, in 1935 the Swiss chemist Paul Müller set out on a research program to discover a compound with the following properties: 1) it would kill insects quickly and efficiently, 2) it would be relatively harmless to plants and mammals, 3) it would be inexpensive to produce, 4) it must be relatively stable once produced, and 5) it should not have any unpleasant odors associated with it. A substance that met such requirements would be a virtually ideal insecticide; there seemed to be little chance that such a compound could be found. In the 1930s some of the most effective pesticides were arsenic compounds that posed dangers to all living organisms and also accumulated in the soil.

Müller concentrated on chlorine-containing organic compounds, and after four years of research he hit upon DDT in September 1939. DDT seemed to have just the characteristics he was looking for; commercial production began in 1942 and its first big test came in 1943-1944. World War II was in progress and the English and American troops had just captured Naples when a typhus epidemic broke out. This was a very serious matter because a major epidemic among the Allied forces could stop them in their tracks as they fought the Nazis; some even worried that it could turn the war against the Allies. Typhus is spread only through the bite of the body louse, so if the lice are killed then typhus cannot spread. Under wartime winter conditions lice spread rapidly from person to person, especially since they could hide in thick layers of clothing, bedding, and so forth. To stop the lice, and therefore the typhus, the soldiers and population of Naples were dusted with DDT powder. It worked perfectly! Before the end of the war DDT was being used to stop insect-borne diseases in Japan and other areas. The people to which the DDT was applied appeared to suffer no harmful effects. For his discovery Dr. Müller was awarded the 1948 Nobel Prize in medicine and physiology.

After World War II DDT was used extensively for both agricultural purposes and to control insect carriers of diseases. It is credited with saving millions of human lives by limiting the spread of such diseases as typhus, malaria, and yellow fever. It also became a cornerstone of the Green Revolution which resulted in dramatic increases in food production during the 1950s through 1970s. It became perhaps the most widely used pesticide ever invented. Huge quantities of DDT were used around the world and DDT residues could be detected in virtually all animal and human tissues. In 1970, during the heyday of DDT use, average residue levels of DDT in human fat was 7.95 ppm [parts per million].

By the 1950s it was becoming clear that DDT was not as safe to humans and other mammals as initially thought. Because of its solubility characteristics, DDT accumulates in fatty tissues (pure DDT is a white, crystalline solid that is insoluble in water but readily dissolves in non-polar substances like fats or benzene). It was discovered that DDT causes reproductive problems in birds (manifested most clearly as thin-shelled eggs that readily break) and other animals. Decreased fertility, changes in the sex organs, and alterations in metabolism have all been associated with DDT exposure. When ingested directly in adequate quantities by a human or other mammal, DDT attacks the nervous system causing headaches, fatigue, confusion, irritability, numbness, and other symptoms. Death can occur in one to three days, although if the dose is too low to cause death the victim can recover completely.

DDT, like many pesticides, bioconcentrates up the food chain. DDT applied in agriculture collects in the soil and organisms such as slugs and worms bioconcentrate the DDT which is then passed on to organisms that feed on the slugs and worms, and so on up the food chain--becoming more concentrated at each level. In aquatic systems bioconcentration of DDT may be even more efficient. Many fish accumulate DDT in their tissues at levels that can be hundreds of thousands of times greater than the levels in the water. Sudden declines in fish-eating bird populations have been correlated with their feeding on DDT contaminated fish--examples include ospreys, eagles, peregrine falcons, and brown pelicans. To make matters worse, many pest insects developed resistances to DDT, reducing its effectiveness which caused some users to apply ever-increasing amounts of DDT.

Rachel Carson emphasized the dangers of DDT to both humans and natural ecosystems in *Silent Spring* (1962), the book that many credit with launching the modern environmental movement. After years of controversy, the use of DDT was banned in the United States in 1972 and in many European countries during the late 1970s. It continued to be produced, for export purposes, in the United States until 1976. Today DDT is produced and widely used in many parts of the world, especially developing countries. Indeed, in Latin America the use of DDT has actually increased during the 1990s. As a result of this continued use, DDT remains a common contaminant worldwide. Even in the United States and Canada DDT residues continue to be found on many fresh fruits and vegetables due to traces of DDT that remain in our soils (although these are declining since DDT has a half-life of about fifteen years) and foods being imported from areas where DDT is still actively used.

Questions

1. Many artificial chemicals undergo the following cycle. First the chemical is invented and initially used in a developed country, and is later exported to developing countries. Then the developed country bans the chemical once adverse effects are discovered, but production and use of the chemical continues in developing countries such that the chemical makes its way back into countries where it is banned. This is sometimes referred to as the "circle of poison." How does the history of DDT illustrate the circle of poison?

2. In hindsight, do you think that Paul Müller should have received the Nobel Prize for his work with DDT? Should he be judged for his accomplishments at the time, or should he be viewed from our perspective?

3. Do you think that DDT should have been tested more thoroughly before being put into production and widespread use? What mitigating circumstances were there that might have caused scientists to rush it into use? How much was known about synthetic organic pesticides in general in the 1930s and early 1940s?

4. Some people have suggested that the supposed dangers of DDT have been greatly overblown--especially by Rachel Carson--when in reality this pesticide has done much more good than harm, at least from a human perspective. How would you respond to such a suggestion?

Sources

Asimov, Isaac, 1964, *Asimov's Biographical Encyclopedia of Science and Technology*. Garden City, New Jersey: Doubleday and Company.

Carson, Rachel, 1962, *Silent Spring*. Boston: Houghton Mifflin.

Franck, Irene, and David Brownstone, 1992, *The Green Encyclopedia*. New York: Prentice Hall General Reference.

Harte, John, Cheryl Holdren, Richard Schneider, and Christine Shirley, 1991, *Toxics A to Z: A Guide to Everyday Pollution Hazards*. Berkeley: University of California Press.

-31-

Computer Chips, Carcinogens, and Health Risks

Many people think of the computer chip manufacturing industry as being the ultimate when it comes to "clean" industries. Fabrication rooms where the silicon-based chips are actually handled and treated are commonly called "clean rooms" because extreme care is taken to make sure that no contaminants enter the rooms. The workers typically wear full-body "bunny suits" made of polyester along with goggles and rubber gloves. Furthermore, the workers must take "showers" of forced air to make sure that they are dust and contaminant free. Often many of the actual chemical applications to the chips are done robotically. Given all the precautions taken, everything may seem very safe in the chip manufacturing industry despite the fact that some of the chemicals used in chip manufacturing and preparation are highly toxic, often carcinogenic substances. Indeed, the Semiconductor Industry Association contends that there has never been a known case of a worker being exposed to a carcinogen and developing health problems.

Since the early 1980s, however, there has been concern among some groups that the chemicals used in chip manufacturing may indeed be adversely affecting workers' health. Commonly used chemicals include such highly toxic substances such as arsenic, nickel, and methyl chloride and numerous complex compounds. The elaborate safety measures in a typical "clean room" are not for the safety of the human workers, but for the safety of the expensive silicon chips. A small particle of dust during manufacturing can ruin the chip. Former fabrication room workers have reported that during some operations (at an International Business Machines [IBM] plant in East Fishkill, New York) dangerous chemicals would sometimes splash from equipment and alarms designed to detect chemical leaks would often be ignored when they went off. There is no doubt in these workers' minds that they were exposed to large quantities of dangerous substances.

Since the 1980s, when some preliminary studies suggested a link between workers in chip manufacturing operations and abnormally high rates of miscarriages, groups such as the Silicon Valley Toxics Coalition have been warning of the potential dangers to workers in the industry. Generally such warnings have gone unheeded. Now, however, more and more cases are coming to light and in March, 1996, a lawsuit was filed in the New York State Supreme Court in Manhattan claiming that seven workers from the East Fishkill IBM fabrication plant have suffered illnesses (and in one case, death) from chemicals they were exposed to during the production of computer chips. This lawsuit is the first of its kind--never before have such apparently well-documented cases been brought forward.

One participant in the lawsuit, now suffering from testicular cancer, believes he and several other members of the lawsuit were repeatedly exposed to a chemical compound called Positive Photoresist that carried a warning label stating: "May cause damage to blood forming tissues. May cause damage to testes. In laboratory animal studies . . . birth defects and adverse effects on pregnancy have been observed." Another worker named in the lawsuit has suffered from brain tumors since working in the IBM chip-manufacturing facilities.

Perhaps the most compelling case for a link between cancer and the chemicals used in computer chip manufacturing is the story of three workers named in the lawsuit: Miriam Nicole Sanders, James Gibbons, and Glenn Haight. These three began working together at the IBM facility in 1988; all were young, in their late teens or early twenties, and in good health. Four years later Sanders was dead as a result of a rare cancerous tumor in her colon at the age of only 22. Gibbons developed a testicular tumor and suffers from colitis and bleeding ulcers. Haight is fighting with cancer. These workers and their families believe that it is not just coincidence that they began suffering from serious health problems after working together at IBM--they blame the chemicals they were exposed to. The lawsuit they have filed is not directly against IBM, but rather names the four companies that manufactured the chemicals alleged to have caused their health problems: Union Carbide Corporation, Eastman Kodak Company, J.T. Baker Chemical Company, and KTI Chemical Corporation.

Questions

1. Do you believe there could be a link between cancer and other adverse health effects and the computer-chip manufacturing industry? How difficult is it to prove that a certain person got cancer as a result of working in a "clean room," especially given the long time delay before the onset of many cancers?

2. How would you feel if you knew that workers suffered from cancer or other health problems as a direct result of manufacturing the chips that went into your computer? Many computer-chip workers take the jobs because of the relatively good pay, security, and working conditions; they may have little or no knowledge of the health risks involved. Before starting work, should they be informed of any potential health risks? How desperate for a job would you need to be to risk your good health by working in a chip manufacturing facility?

3. Based on the evidence so far, do you think federal, state, and local government should impose stricter safety codes on the computer-chip manufacturing industry? Will this be easy, or even possible, given the powerful lobby of some computer manufacturers? Might such actions simply drive chip-manufacturing out of the country to areas where safety

requirements are less stringent? If this is the case, what happens to the workers who rely on their jobs with the computer-chip manufacturers?

4. Even though IBM was not named directly in the lawsuit discussed above, what moral or ethical responsibility do you think IBM might have for its workers?

Source

Glaberson, William, with Julia Campbell, 1996, "Ailing Computer-Chip Workers Blame Chemicals, Not Chance." *The New York Times* (March 28, 1996), pp. B1, B6.

-32-

Are All Risks Equivalent?

One of the arguments for the large-scale use of nuclear power posed by some advocates is that nuclear power poses a much smaller risk of property damage and loss of life than do conventional forms of power and many other activities that we take for granted (such as driving automobiles; there are approximately 50,000 deaths due to automobile accidents per year in the United States). But is all risk equivalent? Mathematically, perhaps, equal numbers or probabilities of equivalent outcomes (such as deaths) may be equivalent, but from the point of view of psychology and social acceptance this is usually far from the case.

From the point of view of many people, risks involuntarily forced upon them are thought of in quite a different light from self-imposed risks. Thus people may voluntarily decide to smoke a cigarette or drink a diet soda containing a carcinogen, knowing full well that they risk contracting cancer by these actions, yet the same people may vehemently oppose a nuclear power plant being built in their backyard because they fear that the nuclear plant will involuntarily subject them to an increased cancer risk, no matter how small that risk may actually be. Furthermore, they may not only be concerned about themselves, but also their children and future generations. Smoking the cigarette or drinking the diet soda, in contrast, may significantly increase their chances of disease, but it will have little direct effect on future generations.

Individuals and society also tend to discount future risks. When a 20-year-old puffs on that cigarette today, the risk of damage in the far future (perhaps thirty years in the future) seems remote and not worth considering. Likewise, when a new technology is first developed the perceived immediate benefits often seem to outweigh any negative consequences in the future--furthermore, by the time the future comes, some people reason, a technological solution will have been developed to deal with the negative aspects. Thus during the history of the nuclear power industry, the enthusiasm for nuclear power in the late 1950s and early 1960s overlooked the long term problems that would inevitably have to be faced--what to do with the toxic, radioactive waste that was being generated, how to deal with plants when it came time to decommission them, what the social consequences would be of a large-scale accident, how to deal psychologically and otherwise with people's not unfounded fears of radioactivity leaking from plants and the fuel cycle, and so forth. Now that nuclear power has been with us for over a generation, these problems and risks are manifesting themselves and must be addressed.

Most people tend to be more comfortable with the known than the unknown, even when the known is much more dangerous than the relatively unknown (but how can one know that if the unknown is in fact *unknown*?). In this respect, in the realm of commercial power, nuclear plants have a big disadvantage over conventional fossil fuel burning plants and hydroelectric plants. For hundreds of years we have been used to the dirty soot from coal burning, and the fumes of burning petroleum products. This pollution may be inconvenient and damaging, even life threatening, but at least we are familiar with it. In the case of hydropower dams, we know what a flood is and feel we can deal with it--indeed if the dam is big and strong enough then there should be no problem. In contrast the potential radiation from nuclear power plants is poorly understood by the public, and what is worse, it cannot even be seen, felt, or detected by ordinary means. It is invisible and people feel helpless to deal with it directly. They must take the word of public officials, whom they may not trust, that dangerous levels of radioactivity are not escaping from the power plant.

Mathematically two risks may be equal, but in most people's judgments they will be very different. For instance, two commercial power technologies may each have a risk of one death every two years. Technology number 1 arrives at this risk because once every ten years, on average, a relatively small accident happens and typically five people are killed. Technology number 2 may arrive at this risk by being "perfectly safe" for 20,000 years, on average, before a huge accident occurs in which some 10,000 people are killed. In both cases the risk is one death every two years, over the long run, but most people would pick technology number 1 over number 2 for very good reasons. With technology number 2 the huge accident could occur any time during the 20,000 years, even in the first year. Also, people tend to be more averse to large, concentrated disasters rather than small numbers of deaths spread out over long periods of time (and generally also spread out over large geographic areas). Relatively diffuse, random deaths are easier to accept--less real, except to those who know the individuals who die--than are huge, widely-publicized disasters. Large disasters can cause widespread psychological and social trauma generally unassociated with dispersed, randomized deaths. In our hypothetical case even if technology 1 killed, on average, one or two people a year, and thus had a significantly increased risk over technology 2, the public would undoubtedly prefer to use technology 1.

In the real world, conventional fossil fuel burning plants and the random, dispersed damage they cause are technology 1 of the above scenario, while nuclear power generation is viewed as technology 2. Even if more damage and loss of life has been caused by conventional power plants thus far, many people view nuclear power as simply a very large-scale disaster waiting to happen. Certainly the incidents of Three Mile Island and Chernobyl, as well as the much publicized lack of long-term storage facilities for radioactive waste, reconfirm the general perception of the public in this respect. In the opinions of many, if a potential disaster stands to be large enough (such as killing or injuring tens of thousands, millions, or billions of people), then it is not worth risking no matter how small the risk is. In this view, if a nuclear power plant could melt down and contaminate an area the size of the state of Pennsylvania, and kill hundreds of thousands of people (as has been suggested), then the nuclear power plants should not be built even if the chance of such an enormous accident occurring is virtually zero. Furthermore, the public generally distrusts experts who label certain risks as being extremely low. A risk that has a very low probability of occurring can still occur. Much of the public felt that they were led to believe that a Three Mile Island or Chernobyl type of accident could never occur, yet they did occur. Of course, many people fail to really understand that an even greater disaster may be slowly creeping up on us, a disaster that could potentially kill or adversely affect billions of people around the world, a disaster caused by the continued burning of fossil fuels--namely global warming which will cause, among other things, climatic shifts affecting food production and sea level rises flooding many coastal areas.

Questions

1. In your mind, are risks that are mathematically equal also equivalent? Explain.

2. Explain the difference between voluntary and involuntary risks.

3. Why do many people, especially when they are young, tend to discount the future?

4. Why do people tend to feel more comfortable with the known and less comfortable with the unknown, even if the unknown entails relatively less risk? Why do people tend to fear major catastrophes?

Sources

Ehrlich, P. R., and A. H. Ehrlich, 1991, *Healing the Planet: Strategies for Resolving the Environmental Crisis.* Reading, Massachusetts: Addison-Wesley Publishing Company.

Ross, John F., 1995, "Risk: Where Do Real Dangers Lie?" *Smithsonian* (November 1995), pp. 42-53.

Trefil, James, 1995, "How the Body Defends Itself from the Risky Business of Living." *Smithsonian* (December 1995), pp. 42-49.

-33-

Dealing with Lead Paint

Although the use of lead in paint was banned in the United States in 1978, some 64 million American houses and apartments contain lead paint and 1.7 million children suffer from high levels of lead in their blood (although some of this is the result of sources other than lead paint, such as industrial operations and lead in water pipes). Lead poisoning is primarily a threat to young children (approximately six years old and under, and especially very young children under three)--it can cause a variety of developmental problems, brain disorders, lifelong learning disabilities and loss of intelligence, behavioral problems such as aggressiveness and hyperactivity, and many other medical problems. Children are commonly exposed to the lead in paint by eating paint chips, inhaling dust containing lead paint (generated during house renovations or even when doors or windows rub as they are used), chewing on window frames, putting their fingers in their mouths after touching lead-laced dust, and so forth. For a young child only minute amounts of lead are necessary to cause severe problems.

Some people concerned about the lead paint poisoning of children believe that the way to solve the problem is through lead-paint abatement in all housing units containing young children. Abatement means either removing or covering the lead paint so that it is inaccessible to young children. Such abatement must be done carefully--otherwise it can cause more problems than it solves. During abatement lead dust must not be released, soil outside the building must not be contaminated, surfaces must be cleaned thoroughly, and any lead-containing materials removed from the property must be disposed of properly. Children cannot be on the premises while lead-paint abatement procedures are being carried out.

Certainly once lead paint is totally removed from a property the problem is solved, but the procedure can be extremely expensive--costing $10,000 or more per unit, so a typical three-family house may cost in the neighborhood of $30,000. In some states, such as Massachusetts, lead abatement is required for all dwelling units where young children live. However, many property owners question if full lead abatement measures are really necessary, or if they are simply "overkill" at the expense of property owners. Instead of abatement, one school of thought argues that education is just as effective, or perhaps even more so, in combating lead paint poisoning in children.

Many education advocates suggest that the parents of young children need to take on the burden of responsibility to ensure that lead poisoning does not occur. The house must be thoroughly cleaned on a regular basis to remove lead dust; especially important is cleaning with a wet cloth around windows. The unit should be kept in good repair; peeling paint, for example, should not be allowed to remain. Children should be instructed not to ingest paint chips, they should always wash their hands before eating, shoes should be wiped on a mat before coming inside to reduce the chance of bringing in lead contamination from the outdoors, and so forth. Very young children should not be allowed to crawl directly on the floor and then put their hands in their mouths; they should always be kept on clean blankets. Furniture can be placed in front of or under windows so that the children cannot reach the window sills (a common source of lead-paint contamination in young children). Basically, parents can use common sense to minimize the exposure of their children to lead hazards. Parents should also make sure that their children eat an iron and calcium rich diet because there is evidence that these minerals help prevent children from absorbing lead in their systems.

Abatement proponents agree that education is certainly positive, but argue that in too many cases it is not enough. The parent may simply not care or understand how serious a threat lead is, or despite the best intentions may not maintain the dwelling unit appropriately to minimize the risk of lead exposure to children. In older units occupied by lower income families that money may simply not be there to stop paint from deteriorating, for instance, and landlords can often be quite unresponsive to such problems. It is perhaps unrealistic to expect the parent to continually follow a very young child around making sure that they do not ingest or inhale substances that could contain lead, or to expect the parent to continue to clean areas where lead dust could accumulate.

In the end, many people feel that a combination of abatement and education would be best in an ideal world. But the world is not ideal, and abatement is extremely expensive. In March 1996 the U.S. federal government entered the debate by publishing new regulations (to take effect toward the end of the year) requiring landlords and sellers of properties to disclose to renters and buyers any information known about potential lead hazards pertaining to a particular property. Landlords and sellers must also supply government developed pamphlets concerning lead-paint hazards. Finally, the regulations guarantee that prospective buyers of a property will have ten days during which they can have the building inspected for lead. The regulations do not require that properties be inspected for lead or that lead paint be removed; they are heavy on education but lack required abatement. Concerning these new regulations, Henry G. Cisneros (The Secretary of the Department of Housing and Urban Development) stated, "the hope is that just getting awareness and information about the problem, people will do things that will keep lead problems from being exacerbated." He further noted that the need to remove the danger of lead paint must be balanced with the need for affordable housing, and added that "simply adding a cost to every single [real estate] transaction in the country is too gross a response to something that might require more discreet judgment. . . . We are really out of an era where we can just declare a national solution to a problem this complex."

Questions

1. Do you think it is possible that the threat of lead paint has been overstated, just as many people now contend that the health problems reputedly associated with asbestos were overblown? Even if the problem of lead is not as bad as some suggest, is it "better to be safe than sorry"? In not requiring mandatory lead abatement in all housing units where young children reside, has the federal government "given in" to the landlords, realtors, and others who oppose such measures?

2. Do you think the federal government is doing enough to combat lead paint poisoning in children? Should they perhaps require the abatement (removal or encapsulation) of lead paint in all housing units where children six years old and younger reside, as Massachusetts currently requires? Or are the Massachusetts laws overly stringent? How much responsibility should the government assume in ensuring that children are not subject to potential lead poisoning? How much responsibility should parents assume in ensuring that their child does not ingest lead paint? How much responsibility should the landlord of an apartment assume?

3. In the state of Massachusetts lead hazards must be abated in rental units containing children six and under. This can cost the owner up to $10,000 a unit, or more. The owner cannot legally get around this cost by either evicting families with young children or refusing to rent to families with young children. Do you think this is fair? Should the owners of rental properties be stuck with these costs, or do you think the state should pay for lead abatement measures since it is the state that requires them?

4. Where do you stand concerning the "education versus abatement" debate over lead paint? Is one more important than the other? Or are both education and abatement necessary, at least to some extent?

Sources

Cassidy, Tina, 1996, "Lead Paint: Education v. Abatement." *The Boston Globe* (March 31, 1996), pp. A1, A4.

Gerstenzang, James, 1996, "Lead Disclosure to be Required for All Rentals." *The Los Angeles Times* (reprinted in *The Sun Chronicle [Attleboro, MA]*, March 7, 1996, p. 35).

Meckler, Laura, 1996 "DPA [sic, EPA] to Require Lead-Paint Danger Warning." *The Boston Globe* (March 10, 1996), pp. A1, A4.

Wald, Matthew L., 1996, "Lead Paint: New Rules Announced." *The New York Times* (March 7, 1996).

-34-

The Subtle Dangers of Synthetic Chemicals

The development of literally tens of thousands of synthetic organic compounds--from pesticides to plastics--is one of the hallmarks of the twentieth century. These chemicals shape and influence virtually every aspect of modern life, and because of their widespread use they are now found, even if only in trace amounts, in our soil, water, air, and the living tissue of practically all organisms, including humans. Some environmentalists warn that this soup of synthetical chemicals may be posing significant health risks for humans as well as wildlife. Usually the first health risk that comes to many people's minds is cancer, but in fact the most serious risks, some contend, could involve decreases in fertility and developmental disorders.

The argument is that many synthetic organic chemicals mimic natural hormones, such as estrogen, and when organisms are exposed to even extremely small amounts at critical times in their development these chemicals can produce severe damage, especially in unborn fetuses. Sometimes referred to as endocrine-disrupting chemicals, such artificial substances have been recently blamed for effects including a reported 50% reduction in human sperm counts around the world over the last half century, reduced levels of intelligence, increases in violence, hyperactivity, and birth defects, an epidemic of prostate and breast cancers, and even increasing neglect of children on the part of their parents. Such assertions are extremely controversial, however, and there is no consensus among researchers as to the strength of the evidence, the validity of many studies, how much of an endocrine-disrupting chemical may produce significant results, or whether studies performed on laboratory animals or wildlife are also applicable to humans.

Proponents of a link between endocrine-disrupting chemicals and all sorts of health problems can cite many studies. A classic example is a study of the alligator population in Lake Apopka, Florida. Significant quantities of the pesticide dicofol were spilled into the lake and several years later it was found that 60% of the male alligators had undersized penises and only about 25% of the normal level of testosterone. The conclusion was that the pesticide was affecting their hormonal levels, essentially "feminizing" the males. Dozens of other studies, from Lake Michigan minks to Baltic Sea fishes, have purported to show similar results--including reduced penis sizes in males, "feminized" male behaviors, female sterility, and other problems. Joint Canadian-American studies around the Great Lakes region in the late 1980s found numerous physical deformities in wildlife, as well as behavioral changes, especially in predators such as otters, turtles, and various birds. Seemingly healthy parents would lay unhatched eggs, or produce sickly offspring. The problems were generally attributed to toxic chemicals accumulating in the parents' bodies and thence affecting the offspring.

Certain researchers suggest that the wildlife studies can be validly extrapolated to humans; others disagree and view such extrapolations as misleading. Even among humans, however, there are documented cases of synthetic chemicals causing fertility and developmental problems. Perhaps the best documented example involves DES (diethylstilbestrol), an artificially synthesized mimic of estrogen. DES was given to many pregnant women in the 1940s through 1960s to prevent miscarriages. Evidently it was discovered that DES can interfere with the development of the fetus, particularly causing female sterility and other fertility problems, as well as a rare form of vaginal cancer in young adulthood, in the children of mothers who took DES.

Another already "classic" study is the paper published in 1992 reporting an approximately 50% drop in human sperm counts around the world between 1938 and 1991. The suggestion has been made that this decline correlates, and therefore was mostly likely caused by, the rapid increase in production and use of synthetic chemicals. But not all scientists agree that the results of the 1992 paper on sperm counts are accurate. Opponents argue that reliable statistics on sperm counts are simply lacking. The 1992 study, critics note, combined data from different areas at different time periods, failing to take potential geographic variations in sperm counts into account. They cite more recent studies indicating that although for unknown reasons New York City tends to have abnormally high sperm counts compared to Los Angeles and Minneapolis, for instance, sperm counts have remained relatively constant for several decades in New York, Minneapolis, Los Angeles, and Seattle.

Relative to the supposed increases in violence and hyperactivity, and decreases in intelligence, there is little beyond the anecdotal level to support such trends, critics contend. To then attribute such supposed changes in human behavior to the widespread distribution of synthetic chemicals is unfounded extrapolation. As for the current "epidemic" of breast and prostate cancer, which some suggest is caused by synthetic chemicals, critics suggest that there is no epidemic at all. Instead, the "epidemic" may simply be a reporting effect. New tests can detect instances of breast and prostate cancer that would have gone totally undetected in past decades, plus heightened awareness of such health problems has resulted in more complete reporting of instances of cancer.

Still, the debate continues. Certainly there is overwhelming evidence that certain synthetic organic chemicals (for instance, many pesticides) are dangerous to humans and wildlife in high enough doses, and there is no definitive proof that such chemicals in low doses are not harmful. Some scientists suggest that presently in most contexts the amounts of synthetical chemicals that humans and wildlife are exposed to are so minuscule that any effects must be trivial, especially compared to naturally produced substances. Many plants create natural "pesticides" and hormonelike substances--and people consume such substances in large amounts as part of an average diet. Researchers on the other side of the fence counter that humans and wildlife have probably evolved mechanisms to cope with these natural potential carcinogens and hormone-disrupting chemicals, but we are unable to cope with the newly developed synthetic chemicals with which we are now assaulted. This hypothesis may sound plausible, but it is far from demonstrated.

Questions

1. Had you previously heard about the 1992 study purporting to demonstrate that human sperm counts have dramatically declined since World War II? If so, how much faith did you place in it? What are your current thoughts on the subject?

2. How great is your concern over synthetic chemicals in the environment? How about in the food you eat? What would make you more concerned than you currently are? Do you think such issues need to be addressed more forcefully than they currently are by government and/or industry?

3. One large group of synthetic organic chemicals are the organochlorines--including chlorinated pesticides (such as DDT), PCBs, dioxins, chloroform, CFCs, vinyl chloride, carbon tetrachloride, and so forth. While organochlorines have often been considered the mainstay of the chemical industry and chlorine is even used to kill pathogens in drinking water, many chlorinated organic compounds are known to be carcinogenic or otherwise toxic while most are suspected of having harmful effects. In 1991 Greenpeace launched a campaign to phase out the industrial use of chlorine. Do you think such a complete chlorine ban is called for? Should organochlorines perhaps be evaluated one by one for harmful effects and then their use restricted if necessary? How practical can such an approach be given that there are over 10,000 different organochlorines being manufactured? Phasing out the use of chlorine in most industrialized processes could cost tens of billions of dollars. How well documented should the potential harmful effects of organochlorines be to justify such an expense?

Sources

Allen, Scott, 1996, "Are Chemicals Endangering the Unborn? Provocative Zoologist Says They're Disrupting Reproduction." *The Boston Globe* (March 18, 1996), pp. 25-26.

Amato, Ivan, 1995, "The Crusade to Ban Chlorine." In *Taking Sides: Clashing Views on Controversial Environmental Issues (sixth edition)* (edited by Theodore D. Goldfarb), pp. 132-141. Guilford, Connecticut: Dushkin Publishing Group.

Colborn, Theo, Dianne Dumanoski, and John Peterson Myers, 1996, *Our Stolen Future*. New York: Dutton.

Hertsgarrd, Mark, 1996, "A World Awash in Chemicals." *The New York Times Book Review* (April 7, 1996), p. 25.

Kolata, Gina, 1996, "Chemicals that Mimic Hormones Spark Alarm and Debate." *The New York Times* (March 19, 1996), pp. C1, C10.

Kolata, Gina, 1996, "Sperm Counts: Some Experts See a Fall, Others Poor Data." *The New York Times* (March 19, 1996), p. C10.

Thornton, Joe, 1995, "Chlorine: Can't Live With It, Can Live Without It." In *Taking Sides: Clashing Views on Controversial Environmental Issues (sixth edition)* (edited by Theodore D. Goldfarb), pp. 120-131. Guilford, Connecticut: Dushkin Publishing Group.

-35-

The Controversy over Water Fluoridation

Fluorine is a highly reactive, yellow-green gas that can be extremely dangerous when in the form of hydrogen fluoride gas or hydrofluoric acid, causing severe burns and other damage to humans. However, since 1945 fluorine, in the salt form sodium fluoride or as hydrofluosilicic acid, has been intentionally added to many public water supplies. Sodium fluoride can also be purchased in tablet, gel, and liquid forms, and is found in some toothpastes. Other forms of fluoride, such as sodium monofluorophosphate, are also found in some toothpastes.

The reason that fluoride is added to some water supplies (currently about 60% of all U.S. public water supplies contain fluoride) and toothpastes is that fluoride in small amounts decreases the incidence of dental caries (cavities). As a result of widespread fluoridation, cavities in American schoolchildren have declined dramatically (by some 36% between 1980 and 1990), and many children have never had a single cavity. However, for over fifty years the fluoridation of drinking water has remained controversial. Fluoridation is generally controlled at a local level, and can become a very contentious issue. Although proponents, such as many dentists and health officials, maintain that the practice is perfectly safe, critics attribute all sorts of ills to the fluoridation of public water supplies. Fluoride in adequate quantities is a poison--it was once used in rat poisons, and 5 to 10 grams (0.175-0.35 ounces) can kill an adult. Links between fluoride and cancer, hip fractures, AIDS, birth defects, heart disease, premature aging, and allergic reactions have also been suggested, although not definitively proven.

Indeed, according to a 1990 U. S. Public Health Service review, there is no evidence that fluoride causes cancer in humans, although one study did link a rare form of bone cancer in four rats (out of a test group of 130) to elevated concentrations of fluoride in their water (25 to 100 times the amount found in most fluoridated drinking water). High concentrations of fluoride have been linked to other types of bone problems. Long term (on the order of decades) exposure to high levels of fluoride (several times legal limits) can result in a condition known as skeletal fluorosis, irregular bone deposits which may lead to pain and crippling of the joints. Other studies have suggested that fluoride may be beneficial in treating or preventing osteoporosis.

When hydrofluosilicic acid is added to water as a fluorine source it makes the water slightly more acidic, which may make it more likely that lead will leach into the water from pipes and lead pipe joints. The problem can be avoided, however, by neutralizing the acid with sodium hydroxide.

The most common negative effect of fluoride is mottling of the teeth. If children under about eight years old are exposed to too much fluoride, whether from the food they eat, drinking fluoridated water, brushing with fluoridated toothpaste, or by other means, the teeth may develop streaking, staining, and mottling of the enamel surface. In mild to moderate forms, as it is usually encountered, such mottling is thought to be only a cosmetic effect--the strength or longevity of the teeth is not affected.

According to proponents of drinking water fluoridation, the optimal level to prevent cavities, without excessive side effects, for water supplies is about 1 ppm (parts per million). At this level less than 10% of children will develop mild or moderate tooth mottling. The federal limit for fluoridation of public water supplies, determined by the EPA, is 4 ppm. At such a high level perhaps 30% of all children will experience some tooth mottling. However, less than 1% of the general U.S. public is exposed to water with greater than 2 ppm fluoride, and public water utilities are required to notify all customers if the fluoride level exceeds 2 ppm.

Critics of fluoridation programs contend that the general public should not be submitted to fluoridation against its will. The main beneficiaries of fluoridation are young children; if desired, parents can supplement the fluoride in a child's diet with various pills or fluoride-fortified salt. Fluoride can have ill effects due to high exposures over long periods of time, and it is virtually impossible to control one's exposure to fluoride if the drinking water has fluoride added. Furthermore, due to the widespread use of fluoridated water in preparing processed foods, people today are generally exposed to increased levels of fluoride simply due to the foods they eat compared to fifty or more years ago. Some researchers have suggested that fluoride levels in public water supplies could be greatly reduced without losing the major dental benefits that fluoride has to offer. Indeed, due to the concentrations of fluorides now found in many processed foods, many people may actually be receiving more than the amount of fluoride optimal for anti-cavity purposes. Finally, there is the matter of cost. Adding fluoride to public water supplies can typically cost about fifty cents per person per year. A community of a million people can save $500,000 a year by not fluoridating their water, while those who want the benefits of fluoridation can purchase commercial fluoride supplements.

Questions

1. In your opinion, what are the strongest arguments for the fluoridation of public water supplies? What are the strongest arguments against it?

2. Do you think that perhaps fluoridation of public water supplies was once a good idea, but its time has passed?

3. Should it be the business of local governments to watch after the dental health of young children? If you answered yes to the last question, is fluoridating water the way to do this, or are there other approaches?

4. Are the rights of a citizen who wishes to have fluoride-free water compromised when a municipality decides to fluoridate the water? If you were a member of a city council grappling with the issue of water fluoridation which way would you vote? How would you justify your vote to your constituents?

Sources

Harte, John, Cheryl Holdren, Richard Schneider, and Christine Shirley, 1991, *Toxics A to Z: A Guide to Everyday Pollution Hazards*. Berkeley: University of California Press.

Mohl, Bruce, 1996, "Clenched Teeth over Fluoride." *The Boston Globe* (March 19, 1996), pp. 29, 32.

-36-

The Dangers of Wood Smoke

Wood fires are popular, both for heating and for pure recreation. In certain areas wood can be easily obtained and burned sustainably (wood is burned no faster than it is produced by growing trees), and many people thoroughly enjoy an evening around an old fashioned campfire or find it cozy and romantic to sit by the fireside. However, studies presented at the Society of Toxicology's 1995 annual meeting held in Baltimore strongly suggested that exposure to the smoke associated with wood fires may be bad for one's health.

Wood smoke contains numerous toxic substances, including known carcinogens, such as polycyclic aromatic hydrocarbons, aldehydes, carbon monoxide, and tine organic particles. Based on epidemiological studies in children, wood smoke has been implicated in increasing respiratory illnesses. Controlled studies on mice and rats have confirmed such associations. In one study, carried out by Environmental Protection Agency scientists, a group of mice was exposed to wood smoke for six hours, a second group was exposed to the emissions from an oil furnace, and a third group (the control group) was not exposed to any type of smoke or emissions. All of the mice were then exposed to an air-borne bacterium which causes respiratory infections. After six weeks only 5% of the mice in the control group and in the group exposed to oil emissions had died of the infection, whereas 21% of the mice exposed to the wood smoke had died. Independent studies undertaken at New York University School of Medicine using rats exposed to wood smoke and respiratory pathogens (such as the bacterium *Staphylococcus aureus*) showed similar results. Based on such data, the researchers are convinced of the potential health threats associated with breathing wood smoke.

Questions

1. Based on the studies cited above, are you worried about breathing wood smoke? If you've always enjoyed sitting by a wood fire, will you enjoy it any less the next time?

2. Do you think it is valid to conclude that wood smoke is bad for humans based only on

mice and rat studies? How does the epidemiological data on children (especially one study that linked increased incidences of respiratory illnesses in preschoolers) strengthen the applicability of rodent studies to humans?

3. Do you think that a government agency, such as the Environmental Protection Agency, should regulate wood fires for health reasons? Should it be required that woodstoves be inspected for leaks that could cause people to be exposed to undue amounts of wood smoke? Should open fireplaces be banned? Should fireplaces be banned only in residences containing young children? Or should people have the "right" to expose themselves to wood smoke if they so desire? How does this compare to the issue of smoking cigarettes?

Source

Stone, Richard, 1995, "Wood Smoke Fires Infections." *Science* (24 March 1995) 267:1771.

-37-

The Potential Dangers of Electromagnetic Fields

Modern technological society is heavily dependent on electrical devices: light fixtures, television sets, computers, refrigerators, electric heating, and so on. Overhead and underground power lines criss-cross the land, electricity runs through homes and other buildings, and virtually everyone is exposed to electric and magnetic fields of one form or another.

Electric fields are produced by the presence of electrical charges, and magnetic fields are produced by the movement of electrical charges. Electromagnetic fields oscillate with various frequencies, are characterized by various wavelengths, and carry different energy values. We perceive certain electromagnetic wavelengths as "light" (the different colors of the spectrum correspond to slightly different wavelengths); other wavelengths we feel as heat. Still other wavelengths are referred to as X-rays, gamma rays, radiowaves, microwaves, and so forth. X-rays and gamma rays are very high frequency (and therefore high energy) forms of electromagnetic radiation that are often referred to as "ionizing radiation" because they have the ability to remove electrons from electrically neutral atoms. There is no doubt that X-rays and gamma rays can harm organisms if the doses are adequate; atoms are ionized which can cause tissue damage. But what about low-frequency, non-ionizing electromagnetic radiation as is given off by electrical appliances, power lines, and currents found in the walls of homes and offices? Does exposure to these forms of radiation pose a health risk? This has been a topic of heated controversy since the late 1960s.

The human body (along with the living tissues of any other organism) naturally contains electrical charges and magnetic fields. Voltages can be measured during muscle use, for instance, and there are electrical fields found across cell membranes. It has been demonstrated that small external electromagnetic fields can affect the internal fields found in the human body; for example, certain external electromagnetic fields have been shown to affect the human body's "biological clock." Yet the mere presence of such effects does not prove that small, low-level electromagnetic fields are harmful to human health.

So what is the bottom line? Are the electromagnetic fields (EMFs) found throughout modern technological society harmful or not? Many researchers would answer that at present equally competent authorities disagree and the data is ambiguous. Some studies have suggested that elevated exposures to common EMFs can either directly cause adverse health effects or worsen health problems which were actually caused by some other factor. Perhaps the most famous/notorious study was one carried out in 1979 that alleged a link between elevated leukemia rates in children and exposure to power lines. Although this study has been widely criticized and disputed, further studies have found at least suggestions (if not conclusive proof) that children exposed to elevated levels of EMFs may be at greater risk for leukemia and cancers of the brain and nervous system than other children. Among adults, the effects appear to be weaker. However, a slight increase in risk of leukemia and cancers of the brain and nervous system has been found in some studies of workers employed at electrical power generating facilities or electronic manufacturing plants. It must be remembered, however, that such cancers are extremely rare to begin with. Even with apparently elevated levels, they remain rare diseases.

There are also suggestions that unborn children may be particularly sensitive to EMFs, yet here too the studies are not definitive. Some research has suggested an association between increased numbers of early miscarriages and the use of electric blankets and electrically-heated waterbeds. One study suggested a correlation between the use by pregnant women of video display terminals that emit high levels of EMFs and an increased risk of miscarriage. However, most computer screens used today do not emit such high levels of EMFs. It has also been suggested that any negative health effects associated with computer use may not even be due to EMFs, but may be caused by other stressful factors (such as posture, eye strain, or ultrasound that may be given off; it has been demonstrated that such ultrasound can have at least some ill health effects, such as headaches).

Even if there is no definitive proof that EMFs are harmful, some people take the attitude "better safe than sorry" and therefore wish to limit their exposure. But given the widespread use of electricity in modern society, limiting one's exposure can be very difficult. Living far from power generation facilities and major electric power lines may not be enough. Indeed, the wiring in a standard home may be more of a potential danger than any external power lines. Some standard appliances can also give off extremely high levels of EMFs--such as electric can openers, hair dryers, electric shavers, and vacuum cleaners--but then one must take into account how often and how long such appliances are used. Non-digital bedside clocks and electric blankets can also be significant sources of EMF exposure in some cases (although newer models of electric blankets are especially designed to reduce potential EMF exposure).

In order to determine one's exact EMF exposure level measurements must be taken. This can be performed by a trained expert with the proper equipment; there are also instruments that can be purchased to measure the strength of EMFs. It must be kept in mind, however, that EMFs can vary greatly over time, depending on what appliances are running, how much current is flowing through wires, and so forth. Furthermore, isolated measurements may be relatively meaningless given that the health significance of EMFs has yet to be agreed upon.

Questions

1. It has been estimated that hundreds of millions of dollars have been spent researching the health hazards of EMFs with no definitive results. Has this been a waste of money? Could

the money have been better spent elsewhere? Much of this money has been spent by utility companies. Do such companies have a vested interest in the outcome of their studies? Whose money are they spending? What moral obligations do they have to protect the public from any potential hazards of EMFs?

2. Given the studies that suggest elevated rates of leukemia and certain other cancers in children exposed to major power lines, do you agree with the Florida judge who in 1989 prohibited school children from playing in a particular schoolyard near major power lines? Do you think schools near major power lines should be relocated? Or should the power lines be relocated? How much evidence of a potential health hazard is needed in order to justify such actions?

3. Given what you know about the potential hazards of EMFs, will you make any attempt to limit your own exposure? What other types of activities do you engage in that might pose an even greater risk to your health? Do you think the advantages of electric power for modern technological society outweigh any potential risks of EMFs?

Sources

Beckmann, P., 1992, "Electromagnetic Fields and VDT-itis." In *Rational Readings on Environmental Concerns* (edited by J. H. Lehr), pp. 253-263. New York: Van Nostrand Reinhold.

Massachusetts Department of Public Health, 1993, *Answers to Frequently Asked Questions about Electric and Magnetic Fields (EMFs) Produced by 60 Hertz (Hz) Electric Power.* Boston: MDPH, Bureau of Environmental Health Assessment, Environmental Epidemiology Unit (150 Tremont Street, Boston, MA 02111).

Naar, Jon, 1990, *Design for a Livable Planet.* New York: Harper & Row.

-38-

Should Nuclear Bomb Testing Be Resumed?

On June 13, 1995, fifty years after the dropping of the bombs on Hiroshima and Nagasaki, and at a time when many governments were working toward the decommissioning of nuclear weapons, the newly-elected French President Jacques Chirac announced that a series of live nuclear tests would be carried out by the government beginning in September. Under President Francois Mitterand, the French government had called a moratorium on nuclear testing in April 1992, and it was a shock to many inside and outside of France that testing would resume. Environmentalists from around the world protested, boycotts of French products were organized, over twenty governments lodged formal protests with the French government, and according to polls 60% of the French public opposed the decision.

The sites of the proposed tests were to be Mururoa atoll and nearby Fangataufa, French Polynesia, in the South Pacific. The first test was slated for Mururoa atoll in September 1995, with up to seven more scheduled during the next year or so. Nuclear bomb testing was not new to French Polynesia. Between 1966 and 1974 France detonated 41 atmospheric nuclear bombs above Polynesia, and between 1975 and the 1992 moratorium over 120 underground nuclear weapons tests were carried out, the majority deep beneath the Mururoa atoll. In these tests technicians drill a deep hole (up to a half mile deep) into the volcanic rock that underlies the atoll. The bomb is placed at the bottom of the shaft, then the shaft is filled with concrete and rock debris. Fiber optics are used to detonate the bomb and take measurements in the nanoseconds before it vaporizes. During the explosion most of the force of the blast is contained by the overlying rock and concrete; the immediate bomb cavity collapses and the rock is vitrified (turned to glass) by the heat and pressure of the explosion. From the surface all that is seen is a bubbling, frothing, and churning of the waters in the lagoon above the explosion. Immediately after the explosion another shaft is drilled into the rock to make more measurements and recover radioactive samples for analysis.

French government scientists working on the project claimed that the renewal of nuclear tests would have no detrimental effects on the ecology of the atoll, or on the environment in general. Many environmentalists and scientists outside of the French government are extremely skeptical of these claims, however. Concerns were expressed that, due to past nuclear tests, the Mururoa atoll is already extremely unstable and renewed testing would only exacerbate the condition. Because of the past underground tests many experts believe that the radiation equivalent of a Chernobyl-size reactor is entombed in the rocks below the atoll. The official French position is that the radioactivity below the atoll is hermetically sealed in fused glass, rock, and concrete, but others cite evidence to the contrary. In 1987 oceanographer Jacques-Yves Cousteau and his team found evidence of bomb-caused fissures in the atoll's coral and rocks, radioactive cesium-134 was found in the lagoon, and radioactive iodine-131 was found in the plankton of the area. To Cousteau this evidence suggested that the radioactivity beneath the lagoon is not as well sealed as the French government claims. More blasts, it was feared, could release a massive amount of radioactivity held within the rocks. French government and military officials continued to dismiss such concerns, saying that any radioactive contamination that has occurred so far is infinitesimal and might even be due to other causes, such as fallout from the 1986 Chernobyl accident.

Independent of environmental concerns, there were other reasons that many people opposed the resumption of French nuclear testing--nuclear proliferation being a prime concern. Before signing an international Comprehensive Test Ban Treaty (CTBT) in 1996 that would forbid future live tests of nuclear bombs, France felt it should perfect its bomb techniques so as to possess a strong nuclear deterrent. But, many argued, this is an outmoded way of thinking. With the end of the cold war emphasis should be placed on activities other than building bombs. Furthermore, if France continues to perfect its nuclear capability, what are other countries to think? It is feared by many that other countries, particulary among the developing nations, will feel compelled to develop and perfect nuclear weapons also before signing any treaties. Thus, France's actions could lead to a new round of nuclear proliferation.

The French proposal was also criticized on technical grounds. One French official stated that these final live nuclear tests would be necessary to gather additional information so that in the future computer modelling and simulations would be adequate to replace live testing. Experts outside of the French government doubted that seven or eight more live tests would contribute substantially to the simulation techniques being developed by the French. Furthermore, it was suggested, the French could acquire more data by sharing information with other governments, including the United States. French officials responded that they did not want such full cooperation with America or anyone else; rather, they preferred to protect the secrecy of their bomb building capability.

Questions

1. Summarize the arguments for and against France's proposed resumption of live nuclear weapons testing. How do issues of national sovereignty and protection, environmental contamination, and global nuclear proliferation bear on this subject?

2. Whose statements would you place more trust in: those of French government officials or those of environmental advocates opposed to any nuclear bombs?

3. If you were a French citizen and were given the opportunity to vote in a national referendum on whether a few more live nuclear bomb tests should be carried out before France signed the CTBT, which way would you vote? Why?

Postscript

France resumed its program of nuclear weapons testing on September 5, 1995, when a bomb was detonated under the Mururoa atoll. In the days leading up to the test French navy commandos boarded and seized two Greenpeace ships so as to prevent them from entering the test area. With the explosion of the bomb protests were staged around the world. In Berlin a mob of 12,000 pelted the French cultural center with eggs and tomatoes, 10,000 protestors gathered in Santiago, Chile, and a riot broke out in Papeete, Tahiti. As a result of the riot and accompanying looting, Tahiti's international airport, as well as several other buildings, was badly damaged.

Sources

Kittredge, Clare, 1995, "Nuclear fallout." *The Boston Globe* (September 4, 1995): pp. 53, 54.

Sancton, Thomas, 1995, "Trouble in Paradise." *Time* (September 18, 1995), pp. 85-87.

-39-

Methyl Bromide and the Ozone Layer

The potential deterioration of the stratospheric ozone layer has been a major topic of concern among environmentalists since the early 1970s. In October 1995 it was announced that three scientists who did ground-breaking work on how man-made substances can damage the ozone layer had won the Nobel Prize in chemistry: Paul Crutzen of the Max Planck Institute for Chemistry in Germany, Sherwood Rowland of the University of California at Irvine, and Mario Molina of the Massachusetts Institute of Technology. In 1970 Crutzen had first discovered that airborne nitrogen oxides can damage the ozone layer, and in 1974 Rowland and Molina demonstrated that chlorofluorocarbons (CFCs) are extremely destructive to the stratospheric ozone. The work of these scientists led to serious consideration of the importance and fragile nature of the ozone layer. CFCs were banned from aerosol spray cans in the United States in 1978, and in 1987 the Montreal Protocol was established to control world-wide production of CFCs and certain other ozone-destroying substances. Based on later amendments to the Montreal Protocol, CFC production was banned in industrialized nations as of January 1, 1996, and is scheduled to be completely banned in all countries by the year 2006.

But even with the complete elimination of CFCs, will the ozone layer be safe? This is currently a topic of heated discussion. Dangerous levels of CFCs and other ozone-deteriorating gases will remain in the atmosphere for decades, or even centuries, after a complete ban is in place. There are fears among some that CFCs may continue to be manufactured illegally even after being banned, and stockpiles of existing CFCs may become a hot commodity, smuggled from country to country on the black market. Currently CFCs are smuggled into the United States from areas where they are still legally produced. Such CFCs are used in older model car air conditioners and the like. But many authorities do not worry that such CFC smuggling will be a serious long-term problem. Newer cars and appliances use non-CFC coolants, it is not that expensive to modify older models to utilize CFC replacements, and eventually there will be very few older models in existence as they are retired due to obsolescence. Furthermore, CFC are extremely bulky and of low value per pound or volume--professional smugglers can do much better with drugs, for instance.

Serious concern is being raised over other classes of chemicals that may also be contributing to ozone thinning, however. At present the biggest threat posed to the ozone layer, according to some researchers, is methyl bromide. Methyl bromide, a widely used pesticide, is applied to such foods as alfalfa, tomatoes, strawberries, grapes, walnuts and wheat. It has the advantage of killing insects, weeds, and other pests in the soil even before the crop is planted. It is also commonly used to exterminate termites. The problem is that methyl bromide may also be a potent destroyer of the ozone layer. Some studies indicate that a molecule of methyl bromide can break down ozone about 40 to 50 times as quickly as a CFC molecule. It has been suggested that methyl bromide may currently account for 10% of all stratospheric ozone damage.

As a result of such allegations, the U.S. Environmental Protection Agency and agencies of the United Nations have suggested complete bans on the production and use of methyl bromide. Suggested time frames for such bans to take place are 2001 in the United States, and 2010 for the world as a whole.

Whether or not methyl bromide should in fact be banned is quite controversial. Unlike the CFCs, methyl bromide is found in nature--it is not simply an artificial, human-made substance. Indeed, the plankton of the oceans produce prodigious quantities of methyl bromide, perhaps 30 to 60% of all methyl bromide released into the atmosphere according to U.N. estimates. As for the methyl bromide used in agriculture, some researchers (although admittedly funded by the methyl bromide industry) contend that most of the methyl bromide used as a pesticide never enters the stratosphere where it could harm the ozone layer. Rather, it is absorbed by the soil or by the oceans where it may be broken down into harmless substances. Banning methyl bromide, it is contended, would only harm agriculture without significantly benefitting the ozone layer.

Questions

1. Can you draw any parallels between the current controversy over methyl bromide and the controversy in the 1970s over CFCs? (Remember that in the 1970s many CFC industry advocates denied that CFCs could be damaging the ozone layer. At that time it was often argued that ceasing CFC production would hurt the economy and lead to food shortages as refrigeration units, dependent on CFCs, lapsed and allowed food to spoil.)

2. Given that CFCs and methyl bromide are two different classes of chemicals, is it possible that CFCs really do deserve to be banned whereas the dangers of methyl bromide to the ozone layer really have been exaggerated by some researchers? How important is it to point out that methyl bromide is produced in large quantities by natural organisms? Can you trust research that is funded by the methyl bromide industry? How about research funded by the EPA or United Nations agencies that may be biased against any chemical suspected of contributing to ozone layer deterioration?

3. If you had to decide, would you allow methyl bromide to be used unchecked, would you limit but not eliminate the use of methyl bromide, or would you totally ban the production and use of methyl bromide? Justify your position.

Sources

Allen, Scott, 1996, "Saving our Skin." *The Boston Globe* (January 1, 1996), pp. 45, 48.

Anonymous, 1995, "Of Ozone and Fruit Flies." *Time* (October 23, 1995), pp. 82-83.

Valente, Christina M., and William D. Valente, 1995, *Introduction to Environmental Law and Policy: Protecting the Environment through Law*. Minneapolis/St. Paul: West Publishing Company.

-40-

Ultraviolet Damage in Canadian Lakes

Acid rain, global warming, and depletion of the ozone layer (resulting in increased concentrations of ultraviolet radiation on the surface of Earth) have been serious concerns for many years. Some researchers have expressed fears that delicate aquatic ecosystems may be particularly sensitive to these types of onslaughts. A Canadian study suggests that while freshwater lakes may be damaged by any one of these factors, in combination they can have a synergistic effect. In particular, the damage caused by acidification and warming promotes ultraviolet radiation damage even in the absence of stratospheric ozone depletion. With ozone deterioration, the lakes are aggravated even more.

The study, carried out in the Experimental Lakes Area of northwest Ontario, was headed by Dr. David Schindler of the University of Alberta. The Experimental Lakes Area has been intensely studied by Canadian scientists since the 1970s and thus provides a sound, long-term data base on aquatic life in densely forested regions. Studies of the lakes in this area previously demonstrated that phosphate pollutants (such as those in detergents) can cause explosive algal growth, and it was data from the same lakes that helped document the destruction of aquatic systems caused by increased acid concentrations (acid rain/acid precipitation). Schindler and his colleagues extended this work by analyzing the effects of acidification, temperature changes, and ozone depletion simultaneously.

During the 1970s and the 1980s the average temperature of the Experimental Lakes Area increased by approximately three degrees Fahrenheit, and whether this was due to true global warming or simply normal cyclical weather variation in the region, it provided a test case of the effects of warming on lakes. What was found is that the warming resulted in increased evaporation, decreased precipitation, decreased groundwater flow, and most importantly for the aquatic communities, decreased flow of organic matter from the forests into the lakes. Dissolved organic matter in the lakes naturally absorbs ultraviolet light, protecting the organisms below the surface. Schindler's study found that due to the decreased organic content of the lakes, the penetration of ultraviolet light below the surface increased by 15 to 20%.

Independent of regional (or global) warming, it has been long known that acidification of a lake will generally decrease the organic content in the lake water. The acid causes dissolved organic molecules to clump together; the heavy clumps then settle to the bottom and/or are decomposed by bacteria. Either way, necessary organic matter is removed from the water column, increasing the penetration of ultraviolet light.

When warming and acidification are combined, the results are far more severe than either factor alone. Not only do fewer organics enter a lake, but those that do are quickly removed by the acidification process. In one lake examined by the Canadian study ultraviolet penetration increased from one foot below the lake's surface to nine feet. But things can become even worse for some lakes. Due to decreased rainfall and increased evaporation, the rate at which some lakes drained and cycled their water decreased significantly. For example, in one lake during the 1970s it took five years for the water to completely cycle (turn over); now it takes approximately twenty years. This means that any remaining dissolved organic matter in the water is exposed to ultraviolet radiation for much longer periods of time. Essentially this organic matter becomes "bleached" by the ultraviolet light, thus decreasing its effectiveness in blocking ultraviolet light from penetrating deeper into the lake.

Add increased concentrations of ultraviolet light hitting the lakes to the problems caused by warming and acidification, and it is easy to understand that ultraviolet intensities will increase significantly below the surfaces of the lakes. But is this a genuine problem for the aquatic life in the lakes, or can they tolerate increased levels of ultraviolet radiation? This is a topic still being investigated, but already some results are in. The increased ultraviolet light can directly break down dissolved organic particles into simpler compounds, and this can effect the food chain--large molecules that may have been safe from microbial activity, for example, may be decomposed by bacteria after being split by ultraviolet light, thus denying the molecules as a food source for other organisms. Trout and other fishes have shown evidence of sunburn-like damage that has been attributed to increased ultraviolet penetration of their habitats. Research has documented decreases in photosynthetic capacity and growth among various algae, the base of the food chain, due to abnormally high levels of ultraviolet radiation. Other studies have demonstrated damage to various invertebrates in lakes exposed to increased ultraviolet radiation, although at least one study suggested that certain crustaceans might be able to grow pigments to help protect them from increased ultraviolet radiation. The bottom line is that certain increased levels of ultraviolet radiation will affect aquatic ecosystems, although scientists are not yet sure what all the effects will be.

Questions

1. How common is it in nature that various assaults on an ecosystem will have synergistic or multiplicative effects--that is, the combined effects of the assaults are much worse than any of the effects taken in isolation? How often does it happen that various effects "balance each other out"?

2. Were the results of the Canadian lakes study to be expected? Would you have been surprised if they were different?

3. How important are these lakes in and of themselves? What is the bigger story here? Should we expect the unexpected when it comes to interactions of various human assaults on the environment on a global level? For instance, might the combination of global warming, ozone depletion, and worldwide chemical and radionuclide pollution cause unexpected, synergistic effects that would not have been predicted by studying each factor in isolation?

Source

Luoma, Jon R., 1996, "For Lakes, Ultraviolet Danger Doesn't Come Just From Sky." *The New York Times* (February 27, 1996), p. C4.

-41-

The Dump in Wellesley, Massachusetts

Wellesley, a town of about 27,000 people just outside of Boston, has achieved a recycling and recovery rate (including material composted) of over 40% through an entirely voluntary program. Wellesley has an internationally famous Recycling and Disposal Facility that is located on the grounds of a now defunct incinerator (many years ago Wellesley's incinerator had to be shut down because it did not meet federal emission standards). In Wellesley there is no curbside pickup of waste, and residents must either pay a private contractor to collect it or take it themselves to the dump. Better than 80% of Wellesley households transport their own trash to the dump; it is estimated that more than 3,000 cars may arrive at the Wellesley facility on a typical Saturday.

At the Wellesley dump there are various receptacles for separated recyclables--glass, paper, aluminum, and so forth. But what distinguishes the Wellesley dump is the variety of different items that can be dropped off for reuse or recycling. Many different grades of paper are separated and collected. Not only the usual aluminum cans, foil, trays, and steel cans are accepted, but virtually all metal objects--from brass bedframes to copper pipes to car parts--are collected for reuse and recycling. Goodwill Industries has a trailer at the dump for the collection of used clothing, small appliances, and the like. There is a "take it or leave it" section where anyone is free to drop off items that they no longer want (appliances, bric-a-brac) and also take whatever they may be able to use. If items remain in the "take it or leave it" area too long, then they are disposed of. There is also a specific book exchange area where people can leave unwanted books and take any other books that interest them.

In the basement of the incinerator is an area designated for the collection of such items as waste oil, car batteries, and used tires. Not all hazardous wastes are accepted in Wellesley, but the reuse of "leftover" household hazardous wastes (such as paints or thinners) is promoted. To this end there is a bulletin board available for people to post notices about leftover materials that they want to give away, or materials that they are seeking. Such an information exchange helps to reduce the amount of material that ends up in the waste stream (one person's trash is another person's treasure). Leaves, grass clippings, and other yard waste is collected and composted; residents of Wellesley may take the resulting compost. When town-owned trees are cut down the wood is split for firewood and made available to Wellesley residents.

The Wellesley Recycling and Disposal Facility sells what it can of the recyclable materials brought to the dump, and the money goes into the town's general fund. The remaining nonrecyclable waste is shipped to a landfill. Commercial companies may deposit rubbish at the facility for a fee, except for separated recyclables which can be left for free. Between avoiding landfill tipping fees and money earned from the sale of recyclables, it is estimated that the Wellesley dump saves the town approximately two hundred thousand dollars each year. In addition, the dump is the focus of much community activity--many people actually enjoy going there as a place for socialization while taking care of a necessary chore--and there is no doubt that the positive aspects of reuse and recycling fostered by the Wellesley dump go far beyond the money saved by the town.

Questions

1. Why do you think the Wellesley recycling program is so successful? What incentives do the residents have to use the dump?

2. Would you feel comfortable patronizing the Wellesley dump? Why or why not?

3. Should similar "dumps" be introduced in communities across the nation?

Source

Connett, Paul H., 1991, "The Disposable Society." In *Ecology, Economics, Ethics: The Broken Circle* (edited by F. H. Bormann and S. R. Kellert), pp. 99-122. New Haven: Yale University Press.

-42-

Recycling Disposable Diapers

It is estimated that 85% to 90% of all North American babies now wear disposable diapers, resulting in billions of diapers being added to our solid waste problem every year. In many ways disposable diapers have come to symbolize the garbage crisis. They are obnoxious, obvious, and filthy in the most common sense of the word. They are seen as litter on our streets, in our parks, on our beaches; they are an obvious component of our landfills, constituting about 1-2% of the volume of a typical landfill. Disposable diapers are not just a problem in the United States and Canada; they are found in over 80 countries around the world.

It is often assumed that reusable cloth diapers are more environmentally friendly than disposable plastic and paper/wood pulp disposable diapers, but in fact this may or may not be the case. Whereas disposables are used only once and then either incinerated or sent to a landfill, reusable cloth diapers also produce enormous amounts of pollution, and use large amounts of energy and resources (especially water), during initial production and subsequent washing and reuse. Any type of diaper, it seems, is detrimental to the environment. When it comes right down to it, many parents pick disposable diapers simply for convenience.

Now one company, Knowaste Technologies Inc. of Toronto, Canada, is acknowledging that disposable diapers are here to stay whether they are detrimental to the environment or not. With funding from the Canadian government, Knowaste has developed the technology to actually recycle disposable diapers and related products. Knowaste collects used diapers, as well as adult incontinence products and bed pads, from day care centers, hospitals, and nursing homes in Canada. At its Toronto facility workers shred the diapers, pulp and launder the material, and separate it into three main components: about 50% wood pulp, 35% plastics, and 15% super-absorbent gel polymers found inside diaper linings. The wood pulp can be made into other paper products, such as tissues and toilet paper. The plastics can be successfully manufactured into an oil-absorbent product that is used in oil spill cleanups, be they huge spills along a marine shoreline or only minor spills in a home or machine shop. As for the absorbent gel polymers, the company is developing a way to turn them into soil enhancement products that will help dry, sandy soils retain moisture. Finally, more disposable diapers can be made from the products of the recycled diapers. Using their experience with disposable diapers, Knowaste is developing ways to successfully recycle milk cartons, juice boxes, and other products.

In 1995 Knowaste's technologies were offered in the United States for the first time. Dydee Diaper Service, a Massachusetts company that has been in the reusable diaper business since 1933, teamed up with Knowaste to offer the option of recyclable disposable diapers. For over 60 years Dydee would deliver fresh cloth diapers to its customers once a week, the next week picking up the soiled diapers and leaving a clean set. Now a Dydee customer can pick recyclable disposable diapers instead. Dydee delivers the disposable diapers weekly, and collects the previous week's used ones (which are stored by the customer in a special deodorized bin provided by Dydee). The used disposable diapers are then trucked to Toronto for processing by Knowaste, but if all goes well Knowaste hopes to build disposable diaper reprocessing plants in the United States and Western Europe.

Questions

1. Seven years of government-funded research went into developing the technology to recycle disposable diapers. Do you think this was time and energy well spent? Why or why not? Could it lead to even more dramatic techniques for recycling other materials?

2. How would you feel about using tissues or toilet paper if you knew that the same paper pulp had once been part of a used disposable diaper? Are such negative associations an obstacle to manufacturers attempting to market products made from recycled disposable diapers?

3. Do you think Knowaste Technologies has totally solved the problem of disposable diapers? In ten years what percentage of disposable diapers do you predict will actually be recycled? Do you think disposable diapers can be recycled in a closed-loop system indefinitely?

Sources

Dydee Diaper Service, no date, "An Exciting New Choice in Diapering." [Brochure] Dydee Diaper Service, Box 223, Yarmouthport, MA 02675.

Humphrey, Penny, 1995, "Diaper recycling in its infancy in U.S." *Taunton [MA] Daily Gazette* (March 23, 1995).

Maher, Paula, 1995, "Diaper service recycles disposables." *Cape Cod Times* (March 28, 1995).

-43-

The Wastefulness of Reuse and Recycling

Some "authorities" argue that in fact the reuse of goods, and the recycling of materials, is "wasteful." But the question is, "wasteful" of what? Here such people often resort to semantic games. While making new products from virgin materials undoubtedly consumes more raw resources and vastly larger quantities of energy than reusing or recycling, more human labor may be entailed in reuse and recycling--thus reuse and recycling are "wasteful" of human labor. Promulgating this viewpoint, Dr. George G. Reisman, a professor of economics at Pepperdine University's School of Business and Management (Los Angeles, California) has written: "To us, used tin cans, paper wrappings, and the like, which cost us hardly any labor to produce or to replace, are generally not worth the trouble of saving or reusing. In fact, it is usually wasteful for us to do so: it wastes our labor and our time, which are the only things in life we should be concerned about not wasting. For if we can produce new tin cans easily, by scooping iron ore out of the earth in ten and twenty-ton loads, it is simply ludicrous to take the trouble to gather up each little tin can and carry it off to some recycling center, because in doing so we spend far more labor than we save" (emphasis in the original, from Lehr, ed., 1992, p. 635).

Dr. Reisman and others of like mind are clearly of a certain philosophical persuasion that values human labor and time above all other considerations. Yet even if one accepts human comforts and values as the center and measure of all things, one can still cogently argue that reuse and recycling--especially as they entail less energy use and therefore less pollution--contribute ultimately to a better quality of human life. Perhaps the little extra bit of human labor and time spent in sorting refuse for reuse and recycling is not wasted after all. In a different context, more labor-intensive industries (such as some reuse and recycling centers that sort items by hand) are often of value in creating jobs and keeping the unemployment rate low. Perhaps we need to return to a philosophy that values honest human labor in its own right. Perhaps jobs should be viewed as inherently meaningful and important--especially jobs that involve dealing with "waste management." Employees in this sector deserve respect, not simply adequate compensation for spending their time pursuing disagreeable work.

Questions

1. Do you think that reuse and recycling are wasteful? If so, wasteful of what?

2. How can one measure wastefulness? How do we define the term "wasteful"? What standard or measure do we apply?

3. Some people enjoy gardening and growing their own vegetables instead of simply purchasing them from a grocery store. Similarly, might not some people actually receive satisfaction from recycling and reusing materials?

Source

Reisman, George G., 1992, "The Growing Abundance of Natural Resources and the Wastefulness of Recycling." In *Rational Readings on Environmental Concerns* (edited by J. H. Lehr), pp. 631-636. New York: Van Nostrand Reinhold.

-44-

Should Toxic Waste Attract More Toxic Waste?

In the decades after World War II uranium was in high demand as the essential component for nuclear weapons. One of the results was a flurry of uranium mining, including a mining and milling (for the preliminary processing of the uranium ore) operation near Ford, Washington. During the late 1950s and early 1960s massive amounts of uranium ore were processed there by the Dawn Mining Company, but then work slowed as the need for more bombs decreased. But in the early 1970s it picked up again as the nuclear power industry blossomed. In order to dispose of the wastes--the tailings--from the mining and milling, the company dug an enormous hole in the ground in which to dump the low-level radioactive wastes. This tailings pond covers about 28 acres and is up to 70 feet deep. It was expected that such a huge volume would be needed to bury the wastes generated by the mine and mill over the next few decades; however, things turned out differently.

In the late seventies and early eighties the public's confidence in nuclear power faltered. In addition, with the end of the Cold War and the collapse of the Soviet Union, the international uranium market was flooded with high-grade uranium from weapons. The result was a general collapse in the uranium market which forced Dawn to shut down. Now the tailings pond sits 90% empty, and it is estimated that it will cost about $20 million to cap and reclaim the site--$20 million that Dawn says it cannot afford.

Two solutions to this dilemma have been proposed: 1) local citizens sue Dawn Mining Company and in that way force them to pay for capping and reclaiming the area, or 2) allow Dawn to go into the low-level radioactive waste disposal business. Under the latter scheme, Dawn would be paid to import radioactive waste from around the country to Ford, where it would be disposed of in the old tailings pond. Once the hole was filled to capacity, Dawn would use the revenues generated to cap the site and reclaim it and as well as closing down and reclaiming the associated uranium mine.

The people of Ford became polarized over these two options. Some argued that you do not solve a waste problem by importing more waste. These residents didn't want stockpiles of radioactive waste in their community in the first place, and they strongly opposed more being trucked in. As to the argument that Dawn Mining could not afford $20 million to take care of the problem, they point out that actually Newmont Mining Corporation owns a 51% share in Dawn. Newmont is the largest gold producing companies in North America and should be able to afford the necessary remedy. After making enormous profits, some citizens contend, these company refuse to take on the responsibility of cleaning up after themselves. Clearly, many felt, the local residents should not have to tolerate such a situation--the companies should be forced to provide the remedy.

Other Ford residents saw the second alternative, allowing Dawn to import more waste, as innovative, realistic, and pragmatic. They seriously feared that if sued Dawn could go bankrupt and then the local citizens would be left with the waste and no reasonable means to deal with it. At least with this plan there would be some assurance that the problem would be dealt with, and without taxpayers having to underwrite the costs. Officials of Dawn and Newmont also argued that they are not legally obliged to reclaim the tailings pond (a point that some citizens disagree with, but it would probably require an expensive court case to resolve), and they have a responsibility to the shareholders not to spend money unnecessarily.

Questions

1. Some Ford citizens believe that Dawn's arguments and tactics amount to a form of blackmail: allow the company to do what they propose (import more waste) or they will leave, leaving their garbage behind while taking years of profits. Evaluate this argument.

2. Even if allowed to follow its proposed plan, some citizens worry that Dawn may not be trusted to ever properly fill, cap and reclaim the tailings pond and old uranium mine. Given the track record of Dawn and other many other mining companies, do you think this is a realistic fear?

3. Assume that you believe that Dawn is in the wrong and ideally should simply reclaim the tailings pond without importing more waste. If we could see into the future and "know" that Dawn proposal will work (Dawn will successfully fill the tailings pond with waste and properly reclaim it by planting trees and introducing native fauna), would that affect your judgement concerning this dilemma? When must one sacrifice a bit of "moral high ground" for practicality?

4. Taking everything into consideration, which of the two solutions outlined above would you favor?

Postscript

Washington state officials, along with a number of influential Ford citizens, approved Dawn's plans to import more waste and then use the proceeds to cap and reclaim the site. In 1996 Dawn Mining Company entered into the low-level radioactive waste disposal business.

Source

McGrory, Brian, 1996, "Town sees toxic waste problem as solution." *The Boston Globe* (March 1, 1996): pp. 1, 4.

-45-

Liability for Pollution from a Superfund Site

The old, inactive Camp Edwards area military reservation on Cape Cod, Massachusetts--a 22,000-acre base--was designated a federal Superfund site in 1989. This is one of the largest Superfund sites in New England and an estimated 53 billion gallons of underground water have been contaminated by toxic substances. Toxic plumes of ground water migrating from the base now flow below the properties of homeowners in such towns as Bourne, Mashpee, Falmouth, and Sandwich. There was general fear that the state might deem the homeowners responsible for cleaning up the mess under their property, and indeed the wording of the state laws did seem to place liability on the owners rather than the military reservation from which the toxic substances originated.

Understandably potential homebuyers were wary of purchasing homes in the area, banks and lending institutions were very cautious about issuing mortgages due to concern over liability, and current homeowners were afraid that they would be stuck with huge bills to clean up a problem they did not create. Fortunately, in this instance, the federal government did acknowledge that it was responsible for the pollution and as of April 1, 1996 the state of Massachusetts declared that the residents of the area would not be liable for any pollution, even when found on or under their property, created by the military base. It has been suggested that such a "Good Neighbor Policy" should become statewide environmental policy, and perhaps even become a model for environmental liability legislation across the nation.

Questions

1. Is there any doubt in your mind that since in the incident described above the military and federal government were responsible for creating and releasing the waste in the first place, they should be responsible for cleaning it up? Just because the homeowners are declared not liable for the pollution does this necessarily guarantee that the federal government will clean up the pollution, or will it perhaps simply remain?

2. How much liability should a property owner assume when he or she purchases a piece of real estate that was previously polluted? Should it matter whether or not the new owner was aware of any contamination prior to purchasing the property? Could the Good Neighbor Policy lead to abuses if an investor knowingly purchases a badly contaminated piece of property for a moderate price, forces the party responsible for the pollution to clean up the property, and then sells the property at a handsome profit?

3. Even though the state deemed that homeowners and purchasers would not be liable for the pollution emanating from the old military base, real estate agents do not predict that there will be a sudden sales boom in the area. Why not? If given a choice between moving to an area near a Superfund site and another area that is not, which area will most people choose? What types of factors (for instance price, the attraction of the Cape Cod beaches, etc.) might encourage people to move to an area near a Superfund site?

Source

Rosen, Isaac, 1996, "Cape Homeowners Deemed Not Liable for Pollution Costs." *The Boston Globe* (April 2, 1996), p. 19.

-46-

Nuclear Reactors at the Bottom of the Oceans and in Space

A 1989 study undertaken by Greenpeace and the Institute for Policy Studies found that there are at least nine nuclear reactors now lying on the seabed. In addition, at least 50 nuclear warheads have ended up at the bottom of the ocean.

Nuclear reactors are used to power some ships and submarines, and over a thousand accidents have been reported concerning such vessels at sea. A number of ships and submarines have sunk, and their nuclear power plants and any nuclear weapons aboard have sunk with them. In addition, radioactive wastes have been lost at sea, and nuclear warheads have periodically been inadvertently dropped overboard. It is obvious that if these nuclear materials are not already leaking significant amounts of radioactivity (military organizations tend to downplay any current leakages), they eventually will as the vessels and casings of nuclear reactors and nuclear weapons corrode and disintegrate. No one knows how serious the damage will be once the radionuclides are dispersed throughout the oceans.

Nuclear reactors have also been put into space, being used to power satellites. Already there have been accidents: In 1978 a Soviet satellite had problems and ended up dumping radioactive debris over 40,000 square miles of Canadian territory. In 1989 the United States space probe *Galileo* was launched from a space shuttle toward Jupiter. On board *Galileo* were fifty pounds of plutonium fuel for the probe's nuclear reactor. Many nuclear-powered satellites and spacecraft are classified military operations, and so it is hard to know the extent of their existence. Part of the Strategic Defense Initiative (SDI, known popularly as "Star Wars") program included the building of nuclear reactors that will orbit the Earth, and a study undertaken in the late 1980s estimated that there were at least 50 radioactive satellites in orbit. We have been relatively lucky so far, as there have been no major nuclear disasters in space--but perhaps "lucky" is all we have been. A nuclear disaster descending from space may well be just a matter of time. If *Galileo* had malfunctioned, or the shuttle carrying it had exploded (as the *Challenger* shuttle exploded), the fifty pounds of plutonium could have been spread widely over the surface of the Earth causing untold damage.

It has been proposed that nuclear reactors be banned from space; instead satellites and spacecraft could depend on solar and other safe forms of power generation for on-board operations. But such suggestions are not popular with the military since non-nuclear powered satellites are bulkier and thus can be more easily detected by the enemy. The political problems of an international ban on nuclear reactors in space, and the monitoring of such a ban to make sure that all parties conformed, also seem all but insurmountable at present.

Questions

1. Why is it difficult to get accurate information on the number of nuclear reactors that have been shot into space?

2. Does the military of any nation have any motivation to advertise that nuclear warheads or nuclear reactors have been lost at sea? Could such incidents be validly considered classified information? Why?

3. Do you think that nuclear reactors should be banned from space? Nuclear power is used on many submarines. What are the chances that nuclear reactors could be banned from ocean-going vessels in the near future? Is there any compelling need to promote such a ban?

Source

Naar, Jon, 1990, *Design for a Livable Planet: How You Can Help Clean Up the Environment*. New York, Harper and Row, Publishers.

-47-

The Garbage Sifters of Cairo

One of the best ways to accomplish reuse and recycling (best in the sense of the maximum amount put to use again) is through intensive hand sorting of rubbish to salvage usable goods and raw materials. But such an occupation typically carries little social status or earning power; indeed, many "Westernized" people feel that such labor is beneath the dignity of any human being, so attempts to build sophisticated machinery to accomplish the same tasks have been pursued. But in urban areas in developing countries there are typically whole communities of people who make their living from collecting, sorting, and selling everyone else's trash. Certainly people involved in these activities typically endure terrible working and living conditions (often living right in garbage dumps), and are poorly remunerated for their efforts, but they do perform the invaluable service of dealing with the rubbish--a service that in some cases could not be easily performed in any other way.

A case in point are the Zabbaleen, or garbage sifters, of Cairo, Egypt. In this city of nearly ten million the Zabbaleen traditionally take care of the trash, and in the process have been running one of the oldest and largest reuse and recycling operations in the world. Through a network of tens of thousands of people the Zabbaleen collect the rubbish of Cairo and transport it to their settlements on the outskirts of the city. There it is sorted into piles of bottles, paper, bone, types of plastics and metals, and so on. Through their informal networks the Zabbaleen sell anything of any value, no matter how minor the value. Their volume is large, their overhead is low, and they have many established contacts.

At one point Cairo officials decided that the Zabbaleen system of waste management was old-fashioned, unhygienic (which it is) and relatively crude: therefore waste management should be updated in a Western fashion. Compactor garbage trucks were imported and modern facilities to deal with the trash were established. But it was soon found that the entire system was prohibitively expensive to run, the trucks could not always navigate the narrow streets of Cairo, and the end results were less than satisfactory. In the end some of the elements of the modernized system were integrated with the traditional Zabbaleen system of waste management.

Questions

1. Why are garbage sifters like the Zabbaleen not typically found, at least to any great extent, in the more developed countries?

2. Was it right for Cairo officials to attempt to interfere with the Zabbaleen way of life? Even if their standard of living is low, upgrading the garbage collection and sorting system could put thousands of people out of work. Would this be fair?

3. Why were the attempts of Cairo officials to reform the Zabbaleen system not entirely successful?

Source

Kumar, Ranjit, and Barbara Murck, 1992, *On Common Ground: Managing Human-Planet Relationships*. Toronto and New York: John Wiley and Sons.

-48-

Truth in Advertising and Campaigning

In November 1992 the "Massachusetts Packaging Reduction and Recycling Act" appeared on the ballot for state residents to vote on. The purpose of the proposed act was simply to reduce excess packaging of consumer goods and to increase recycling in the state. For several months previously a fierce campaign was waged by both proponents and opponents of the act attempting to influence the result of the vote. In the process many misleading statements were made, especially by those opposed to the recycling act.

A typical claim was that voiced by Professor Jeffrey A. Miron, then Chairman of the Department of Economics, Boston University. In a widely aired television commercial Professor Miron stated that the proposed act would result in "over $500 million in higher consumer prices." As proponents of the Recycling Act pointed out, the claim of higher consumer prices was based on a badly flawed study conducted by Clayton Environmental Consultants (a group that was heavily funded by the plastics and packaging industry). In a nutshell, proponents of the Act charged, the Clayton study took four non-random sample packages (of the thousands of packages used in Massachusetts) and found that it would cost 5 to 15% more to make those packages out of recycled material. The authors of the study then simply multiplied the total value of all packaging in Massachusetts by 5 to 15%, arriving at the increased cost to consumers of $500 million. The Clayton study ignored the fact that many types of packaging already used in Massachusetts would comply with the Act, without further modification. Furthermore, for those types of packaging that did not comply with the Act modifications to meet the Act's requirements would not necessarily cost more.

In response to the claims of Professor Miron and the Clayton study, the economist Robert Stone and Professor Nicholas Ashford (both of the MIT Center for Technology, Policy and Industrial Development) wrote (quoted in an October 1992 information packet distributed by the members of the Recycling Initiative Campaign):

> It is typical for industry to provide inflated or imaginary costs of compliance whenever they are confronted by government requirements or government-induced change. (Take the projections of the dire economic impacts of CFC or PCB regulation.) These industry-funded projections are simply unrealistic, based on faulty assumptions, and have no basis in fact.

First of all, many firms are already in compliance with packaging requirements and will incur no new packaging costs as a result of the Recycling Initiative. Second, some manufacturers will comply with the packaging requirements by reducing the amount of packaging they use by 25% or more -- such compliance will obviously reduce, not increase, packaging costs. Third, some compliance will be achieved by shifting to materials that have acceptable recycling rates -- again, this need not involve any cost increase. Fourth, the substitution of recycled for virgin materials in packaging use does not necessarily cost more.

In addition, compliance with the Recycling Initiative should impose few transaction costs on industry. Manufacturers of goods change their packaging frequently anyway -- usually every couple of years or so. And retailers will develop standard contract language requiring their manufacturers and distributors to assume liability for any and all costs arising from packaging violations, as happens with most other regulatory requirements (e.g., retailers don't themselves verify that their products are in compliance with health standards, they protect themselves contractually).

Most important, the industry projections completely ignore the hundreds of millions of dollars saved annually as a result of avoided disposal costs. These net savings are expected to range from $175 million to $300 million annually. Similarly, the industry projections omit the recycling business that will be brought to the Commonwealth [of Massachusetts] and the stimulative effect it will have on the Massachusetts economy.

Questions

1. Why do you think industry fought the Massachusetts Packaging Reduction and Recycling Act? Why do industry and business often support the status quo?

2. Which side in this argument appears to have been more honest? Which argument would you accept?

3. Which way would you have voted on the Massachusetts Packaging Reduction and Recycling Act? Why?

Postscript

The petition was defeated at the polls.

4. Why do you think the act was defeated at the polls? How do most people respond to the threat of higher prices? Which side probably had more money to spend on advertising?

Source

Recycling Initiative Campaign, 1992, October 1992 Information Packet (Boston, Massachusetts)

Section 4

Social Solutions

-49-

Green Dollars--A Vancouver Experiment

In 1983 Michael Linton of Courtenay, Vancouver Island, Canada, developed a new local money system that proponents hailed as being environmentally friendly, helping to promote local autonomy, self-reliance, and sustainability. The system described below was used from 1983-1989 on a local level in Courtenay (at its height it included about 500 participants and did hundreds of thousands of dollars worth of transactions), but ultimately failed due to organizational problems. However, similar systems have been established elsewhere--reportedly about fifty local systems worldwide so far.

The new money system originated from the recession and high unemployment experienced in the area during the early 1980s. Linton, like many people, was having difficulties making ends meet so he decided to try bartering his services. He was not very successful in this using the traditional economic system based on cash transactions (with related checks, credit cards, and so on), so he ended up inventing a new type of money that would serve his bartering needs. What he came up with was the "Local Employment and Trade System" (or LETSystem) and a new type of local money known as the "Green Dollar."

Under the LETSystem a number of people who wish to trade among themselves get together and agree to abide by the same rules. Each gets an account number, and a central accounting office is set up. Each person compiles a list of their "wants" or "needs" (perhaps having their roof repaired, piano lessons, or work done in their garden) and their "offers" (for example, repairing cars or dancing lessons). The lists of everyone in the system are centrally compiled and then circulated among the participants. LETSystem allows members to barter without trading directly on a one-to-one basis. Mary may see that John is offering roof repairs, yet she has nothing to offer directly to John. She can still contact John and arrange to have him repair her roof for $300 Green (the currency of the system--to be explained shortly). He does the work and she calls the central accounting office and informs them that John is to be credited with $300 Green and she will have $300 Green debited from her account (initially each account is zero, so she may be minus $300 Green--that is, in debt if this is her first LETSystem transaction). Later, John may arrange with Steve to purchase some home-grown vegetables for $25 Green. John calls central accounting and informs them of the transaction, and $25 is removed from John's account and placed in Steve's account. Mary may be offering violin lessons, and perhaps Steve's daughter is interested in trying her hand at this instrument. Steve makes arrangements with Mary for an hour's trial lesson for $20 Green and calls the central accounting office to notify them of the transaction. And so it goes in the LETSystem. Any participant can arrange a transaction with any other participant. On a regular basis (monthly in the case of the Courtenay LETSystem) each participant receives a statement of activity in their account plus a bulletin summarizing activity throughout the system and listing wants and offers among the members.

The Green dollars ($ Green) are totally intangible--they are simply information, a means of accounting. They do not correspond to any kind of physical entity (like dollar bills or gold bullion). Every person participating in the LETSystem is allowed to go into debt, at least up to a certain limit. Such debt is necessary, at least initially, as everyone begins with an account of zero. No interest is charged on such debts, but likewise no interest is paid on credit accounts either. Initially in Courtenay the Green dollars were based on standard Canadian dollars so that $1 Green equalled approximately the value of $1 Canadian.

For a few years the Courtenay LETSystem was extremely successful--so successful, in fact, that the Canadian government took it seriously enough to tax certain LETSystem transactions. Some commercial shops in the local economy began to use both standard Canadian dollars and Green dollars; in some cases locally-produced goods would be sold for Green dollars while imported, commercial goods were sold for Canadian dollars. In other cases Canadian and Green dollars would be mixed in a single transaction: thus a $10 item might be paid for with $9 Canadian and $1 Green.

Proponents of the LETSystem have enumerated a number of its benefits relative to traditional monetary systems:

1) The LETSystem is simple and easy--it works, even while a traditional monetary system (necessary for transactions outside the local community) remains in place.

2) In the LETSystem money is simply information--it does not exist as a tangible entity that can be lost or stolen. Also, advocates of the LETSystem suggest that participants are apt to be less greedy for Green dollars for their own sake. Accumulated Green dollars do not bear interest, and one does not have to pay for going into debt (within reason--if one runs up too much of a debt, or never pays off any of it, other members may refuse to participate with the debtor).

3) Green dollars used by the LETSystem promote decentralization and creativity of people figuring out what they can offer that other people will want. No one person controls the money supply because there is no money supply. Green dollars are created as participants have products or services to offer. The LETSystem promotes self-esteem and self-reliance.

4) The LETSystem is community-oriented and fosters personal transactions, meaningful social relationships, trust, and friendships. It is not an anonymous system. It promotes community unity.

5) Being limited to local circulation among the participants of the LETSystem, Green dollars promote the local economy over the national or global economy. This helps to strengthen community autonomy, self-reliance, stability, and ultimately sustainability. In this manner the Green dollar truly is green--it promotes environmentally-friendly, sustainable activities on the local level.

Questions

1. How practical do you think the LETSystem is? How much does it depend on the honesty and goodwill of the participants? How does it differ from current monetary systems? How is it similar to modern credit and checking systems? Would you participate a LETSystem?

2. Assuming that the participants are honest, what are the advantages and disadvantages of the LETSystem? How does it promote economic goals? How does it promote environmental goals?

3. Do you think the benefits of the LETSystem outweigh the drawbacks? On how large a scale do you think a viable LETSystem could be successful? (Certain proponents have stated that in order to work a LETSystem should probably be restricted to a certain geographic area and contain no more than roughly 5000 people, but this is only speculation on their part.)

Source

Dauncey, Guy, 1988, *After the Crash: The Emergence of the Rainbow Economy.* Basingstoke: Green Print. [Pp. 52-64 are reprinted, with an editorial commentary, IN Andrew Dobson, editor, 1991, *The Green Reader: Essays Toward a Sustainable Society.* San Francisco: Mercury House.]

-50-

Profits or Environmental Responsibility?

Herman Miller, Inc., of Zeeland, Michigan, is a manufacturer of office furniture, office partitions, and related goods. Founded in 1923 by a devout Baptist, D. J. DePree, the company has a long history of attempting to "do what is right," both in the realm of employee-management relations and the company's impact on the environment.

A distinctive and popular piece of the Miller line of office furniture is the Eames chair. Selling for over $2,000, this is a plush office chair made of Honduran mahogany and rosewood with a leather seat. In March 1990 the research manager for Herman Miller, Bill Foley, was undertaking a routine evaluation of the materials used in Miller products when he realized that the use of tropical hardwoods in the Eames chair was contributing to the destruction of rain forests. Foley halted the purchase of such tropical hardwoods by Herman Miller. Once existing company supplies of these woods were depleted the Eames chair would have to either be made of other woods or go out of production. Even if made of alternative woods, it would no longer be a traditional Eames chair. The CEO of Herman Miller, Richard H. Ruch, predicted the demise of the chair.

Herman Miller is environmentally aware in many areas as well as in the use, or nonuse, of tropical hardwoods. Between 1982 and 1991 it slashed the amount of trash it deposited in landfills by 90%. In large part this was accomplished by building an $11 million waste-to-energy heating and cooling plant that saves Herman Miller $750,000 annually in fuel and landfill fees. It also sells scrap fabric rather than burning or landfilling it. In its manufacturing plant Herman Miller reduced styrofoam packaging by 70%, and it replaced the 800,000 styrofoam cups used annually by its employees with 5,000 reusable mugs. In 1991 Herman Miller spent nearly a million dollars on incinerators that destroy 98% of the toxic substances escaping from the areas where wood is stained and varnished. These incinerators exceed the legal requirements found in the 1990 Clean Air Act, and furthermore they may be unnecessary in a few years when nontoxic wood finishing products become widely available.

Herman Miller has paid a price for its environmentally friendly approach, however. As of the end of fiscal year 1991 the company's net profits had fallen 70% from the previous year. Earnings on total sales of $878 million amounted to a mere $14 million.

Questions

1. Do you think that Herman Miller is making the correct decisions? What obligations (ethically, morally, legally) does it have to the environment? What obligations does it have to investors and others involved with the company?

2. As long as Herman Miller is making some profit, does it matter how large that profit is? Could too little profit entail the demise of the company? If Herman Miller failed, what would happen to its thousands of employees? Explain your reasoning.

3. Even if Herman Miller's profits drop dramatically in the short term, by investing in environmentally sound and sustainable practice, might not the company be assuring its preservation in the long-term? Could you support the contention that Herman Miller's actions actually do amount to good business planning? Do you think publicity concerning Herman Miller's environmental attitude might help sales? (Remember to take into account that they manufacture office furniture which is used primarily by other businesses.)

4. If you were the CEO of Herman Miller how would you handle the issues described above?

Sources

Jennings, M. M., 1993, *Case Studies in Business Ethics*. Minneapolis/St. Paul: West Publishing Company.

Woodruff, D., 1991, "Herman Miller: How Green Is My Factory?" *Business Week* (September 16, 1991), pp. 54-55.

-51-

Green Consumerism, or No Consumerism?

North American society is very much a consumer society. Progress is often measured in the number of goods manufactured, bought, and sold. Time-saving gadgets, appliances, convenience items, toys (for adults as well as children) are the manifest trappings of our way of life. Happiness, for many people, is associated with material affluence and consumption. Buying for the sake of buying has been raised to an art form by some consumers; shopping has become a pastime, a means of relaxation. Most items purchased by the average North American consumer are far removed from the true necessities of life; even many foods are totally frivolous snack or dessert items, and are usually processed and over-packaged.

It is only natural that environmentalism--"greenism"--would come to mix with consumerism. In the 1980s and 1990s there has been a faddish trend to be a "green consumer," a consumer who makes environmentally responsible choices when purchasing products and living the affluent North American lifestyle. Books on how to be a green consumer--what products to buy, which companies to patronize--have become bestsellers. A popular book entitled *50 Simple Things You Can Do to Save the Planet* describes how the American consumer can adjust his or her shower and toilet to save water, avoid pet flea collars containing dangerous pesticides, replace incandescent bulbs with energy-efficient fluorescent lights, and so on. All of these suggestions are positive and will help to solve our environmental problems, but clearly even if we all (everyone on Earth) carried them out to the letter, the planet would not be "saved." In fact, if everyone on Earth had modern running water, well-fed pets, electricity, and other conveniences that most Americans consider "necessities," if all people on Earth lived the affluent life of the average North American (even after taking into account the *50 Simple Things* and more), the planet would surely be quickly doomed--or so the radical critics of consumerism contend.

Green consumerism may be more benign than non-green consumerism, but it is not the solution to our environmental problems, according to critics. In fact, green consumerism may even exacerbate and prolong the problems. Green consumers may become complacent; they think they are doing the right thing, so they no longer need to blame themselves for the environmental mess--it is the fault of those who are not green. To make matters worse, many companies in the 1980s and 1990s have "turned green" simply as a marketing ploy. Even if they have no interest in environmental issues, they have learned that by advertising themselves and their products as environmentally and socially responsible they can sell more products and increase their bottom line profits. A company may use recycled materials or natural ingredients in order to attract consumers; in some cases they may even advertise themselves as "green" falsely. Either way, the typical commercial business establishment does best by increasing consumerism (and hopefully the consumers purchase their product and not that of a competitor).

Some environmentalists argue that what needs to be done is to discourage consumerism in all its forms, green as well as non-green. We must focus on reducing our levels of consumption and waste production if we are to save the planet. Green consumerism can even be counter-productive in this respect. If conscientious people truly believe that they are buying the right environmentally benign products they may purchase things they would not have considered previously, or purchase more of certain types of products. Green consumerism, especially as promoted by commercial businesses, can increase levels of consumption which more than offset benefits derived from using green products over non-green products.

Consumerism in any form, critics contend, is not only bad for the environment but also bad for the social and emotional well-being of people. Consumerism engenders dissatisfaction although it seems to promise satisfaction. The consumer is never satisfied; even as ever more is consumed businesses are always pushing "new" or "better" products. One must always keep up with the neighbors (even if those neighbors are simply the characters in an advertisement). Consumerism reduces self-reliance; a classic example is mothers who have no idea how to feed a baby without using prepared baby foods. Consumerism traps people into thinking that convenience items, frivolities, and luxury goods are necessities. Because of consumerism, people have to work harder than ever, spending more hours at disagreeable jobs, to buy more and more things--yet they feel increasingly empty and dissatisfied with their lives. It would be best for both the Earth and the people of the Earth if the classic North American consumer mentality were abandoned.

Questions

1. Is it ethically wrong for a company to turn green--for example, switching to the use of all recycled materials--simply to increase profits, but then advertise itself as concerned about the environment?

2. Do you agree with the critique of consumerism summarized above? Support your answer with specific examples.

3. Even if the consumer mentality, per se, is basically bad for the environment, if people are going to be heavy consumers anyway, isn't it better to encourage them to be green consumers?

4. Honestly assess your own typical consumption habits. Could you do more to be a green

consumer? Could you consume less? Would you be willing to change your lifestyle for the sake of the environment? Have you bought into the classic North American consumer mentality?

Sources

Dobson, Andrew, 1991, *The Green Reader: Essays Toward a Sustainable Society.* San Francisco: Mercury House.

Earthworks Group, 1989, *50 Simple Things You Can Do to Save the Planet.* Berkeley: Earthworks Press.

Jennings, Marianne Moody, 1993, *Case Studies in Business Ethics.* Minneapolis/St. Paul: West Publishing Company.

-52-

Green Markets in Mexico

Some experts in the field of green marketing believe that the market for environmental goods and services, such as waste remediation techniques, pollution control and monitoring equipment, and so forth, is quickly maturing in the United States and Canada--meaning that there will be less and less business in these areas in the future. This is not to say that the need for such equipment and services will not exist, but demand will be at a lower level as appropriate infrastructures are installed, the worst hazards are cleaned up, and damaged environments are restored. In order to keep growing, some environmental businesses are actively expanding into foreign markets. Currently, one of the markets with the most potential for U.S. and Canadian green firms is Mexico.

Green markets in Mexico abound for U.S. and Canadian firms for a number of reasons. First and foremost, Mexico has its share of environmental problems. Mexico City's air, for instances, is among the most polluted in the world. In 1994, according to Mexican government estimates, the market for air pollution control equipment in Mexico was $139.4 million, most of which is currently imported from the United States. Likewise, when it comes to water pollution, Mexico has severe problems. Approximately 13 million urban Mexicans do not have potable water piped into their dwellings, and 22 million people lack basic sewer services. Seventy percent of Mexican domestic waste water and 85% of Mexican industrial waste water is left untreated before being discharged into rivers, lakes, or the ocean. In 1994 the Mexican National Water Commission announced plans to build 174 additional municipal waste water treatment plants and another 100 industrial waste water plants. Then there are problems with hazardous waste sites, environmental contamination, and the like throughout Mexico. As a whole, the Mexican environmental market was estimated at over $2.5 billion for 1995, and it continues to grow.

Because the United States (and Canada, compared to countries on other continents) is situated close to Mexico, American corporations have an immediate advantage when it comes to shipping equipment and personnel to Mexico. Furthermore, the North American Free Trade Agreement gives preference to North American goods when it comes to tariffs. Another major advantage is that American and Canadian goods and services already have a solid reputation in Mexico. Currently over $1 billion worth of pollution control equipment alone has been imported into Mexico from her foreign neighbors. Another major boost to

American and Canadian companies doing business in Mexico should be the 1994 Mexican corporate tax reforms that promise to provide economic incentives to businesses purchasing pollution control equipment that meets certain standards. Now may well be the time for U.S. and Canadian environmental firms to thoroughly explore business opportunities in Mexico.

Questions

1. Do you think there are many lucrative business opportunities left for environmental goods, services, and technologies in the United States and Canada, or have such opportunities already peaked? In order to succeed, must companies in this field start looking toward foreign markets such as Mexico? What will happen to such companies once all the markets (domestic and foreign) become saturated?

2. If you inherited a large sum of money and were looking for a good investment, would you consider investing in a company that markets pollution control equipment in Mexico? Why or why not?

3. Should American or Canadian companies feel any obligation to attempt to address environmental problems in their own countries before entering foreign markets? Do pollution, overpopulation, and other problems in Mexico directly affect the United States? Should Americans take a particular interest in Mexican environmental affairs?

Source

Greshin, Jeremy H., 1996, "Opportunities in Mexico for Pollution Control Equipment." *Massachusetts' Environment* (January 1996), p. 9.

-53-

Aztec Cannibalism

During the fifteenth and early sixteenth century the Aztecs, with their capital at Tenochtitlan (the site of present-day Mexico City), dominated the Valley of Mexico. Aztec society was marked by rigidly enforced class distinctions and a strong military which seemed to be almost perpetually at war with neighboring states. There was also a large and powerful priesthood whose job was to continuously appease and nourish the gods.

According to Aztec theology, bloody human sacrifices were required to nourish the gods. The gods had once sacrificed themselves in order to create the Earth for humans, and now human sacrifice was a way to return the favor. Every day was a struggle for the gods. The Aztec national god, Huitzilopochtli, battled against the forces of darkness each night in order that the Sun might rise again the next morning. To maintain his strength, Huitzilopochtli required human hearts and blood. Hearts and blood were supplied in abundance by human sacrifice. The rest of the bodies were eaten by other humans--the Aztecs practiced cannibalism on a scale that was never matched before or since.

Although a minority of the sacrificial victims may have been criminals or slaves, the majority of those sacrificed were war captives. Indeed, the Aztecs waged war in part simply to capture sacrificial victims. The level of sacrifice can be gauged from the following statistics: An estimated 20,000 captives were sacrificed over a period of four days in 1487 when a new temple to Huitzilopochtli was dedicated. When the Spanish conquistador Hernando Cortez and his followers entered Tenochtitlan in 1519 they found huge towers composed solely of human skulls, and also large racks on which the skulls of victims were laid out (after the brains had been removed and eaten). It was impossible to count the number of skulls in the towers, but one of Cortez's companions counted 136,000 skulls laid out on a huge rack. Best estimates are that at least 15,000 to 20,000 individuals were being sacrificed and eaten annually in Tenochtitlan. At the time Tenochtitlan had a population of perhaps 150,000 to 200,000 people, so if the sacrificial humans were divided evenly (perhaps an unlikely assumption, as those higher in the social hierarchy probably received a larger share), each person received approximately a tenth of a person a year to eat. When the Spanish first entered the city it was noted that there seemed to be no shortage of fresh meat in the markets, but Cortez believed that much of this meat was in fact human flesh.

Human sacrifice and cannibalism was widely practiced in Mesoamerica before the time of the Aztecs, but the Aztecs placed a major emphasis on these practices. An important question is why? Some anthropologists have suggested that it was simply an outgrowth of the Aztec religion and culture; human sacrifice was the ultimate way in which they communed with their gods. But another explanation, one based on environmental considerations, has also been suggested.

By the time of the height of Aztec culture the wild fauna of the Valley of Mexico appears to have been badly depleted. People had lived intensively in the area for thousands of years and there was just not enough natural animal protein to go around. The Aztecs had few domesticated animals, and what they did have could not be raised in sufficient numbers to satisfy the needs of the estimated 1.5 million people living in the Valley of Mexico. Excluding human flesh, the best estimates are that at most the average Aztec could have acquired but two or three grams (0.07-0.105 ounces) of animal flesh a year--less than half the animal protein found in the notoriously food-poor India of the 1970s. Not only did the Aztecs generally lack adequate supplies of animals as food, but they also lacked a vigorous and reliable farming base. Aztec chronicles record numerous crop failures and famines in the fifteenth and early sixteenth century. One of the worst lasted from 1451 to 1456 and was accompanied by intensive warfare and massive prisoner sacrifice. Human sacrifice and cannibalism, already an accepted part of Mesoamerican religion, became a way to acquire and distribute meat to a hungry populace.

Questions

1. Do you think the environmental explanation for the extent of human sacrifice and cannibalism in Aztec society makes sense? Or is this explanation too simplistic? If you accept the environmental explanation, how did Mesoamerican religious precedents set the stage for the Aztec solution to hunger?

2. Could a system of feeding people with human flesh work without a supply of neighboring states from which to take captives?

3. Accounts of cannibalism often accompany histories of major famines (for example, it is recorded that during a major famine in 1318 in Ireland the recently dead were exhumed from their graves and eaten). Do you think it is morally wrong to eat human flesh? Could you imagine a future society where, due to overpopulation, the ingestion of human flesh was generally accepted?

4. When the Spaniards of the sixteenth century first entered the Aztec city they were astounded and horrified by the extent of human sacrifice. In part they used this as justification for destroying such a barbaric, non-Christian culture. In hindsight, did the Spaniards judge too hastily? Were the Spaniards determined to conquer the Aztecs at any rate? How should we judge cultures and societies that are very different from ours? Or should we judge them at all?

Sources

Atmore, Anthony, et al., 1974, *The Last Two Million Years.* London and New York: The Reader's Digest Association.

Harner, Michael, 1977, "The Ecological Basis for Aztec Sacrifice." *American Ethnologist* 4:117-135.

Harris, Marvin, 1979, *Cultural Materialism: The Struggle for a Science of Culture.* New York: Random House.

Ponting, Clive, 1992, *A Green History of the World: The Environment and the Collapse of Great Civilizations.* New York: St. Martin's Press. [Information on the famine in Ireland, 1318.]

-54-

Thoreau at Walden Pond

Henry David Thoreau (1817-1862), an American naturalist, philosopher, and writer, has been a continued inspiration to environmentalists and conservationists of the late nineteenth and twentieth centuries. With his mentor Ralph Waldo Emerson (1803-1882), Thoreau was a leading figure among the New England transcendentalists. Transcendentalists generally believed that intuitive knowledge was superior to knowledge gained through the senses, and many had a strong interest in nature and what could be learned from the wilds untouched by humans. In order to gain a better appreciation for nature, Thoreau lived from 1845 to 1847 in a cabin he built in the woods on Walden Pond near Concord, Massachusetts. From his experiences living there alone came his classic book *Walden, or Life in the Woods* (1854). In July 1846, while living at Walden, Thoreau was briefly jailed for refusing to pay his taxes to the town of Concord. Thoreau argued that the town was supporting the current Mexican War which was, in his opinion, an attempt to support and spread slavery--therefore out of conscience he could not pay his taxes. From this experience came his "On the Duty of Civil Disobedience" which has since become a manifesto for those preaching passive resistance to government injustices--including many environmentalist activists.

Thoreau's fame came after his death. During his lifetime he published only two books, *A Week on the Concord and Merrimack Rivers* (1849) and *Walden* (1854). Neither was commercially successful; indeed, of 1000 copies printed of *A Week*, after four years over 700 were returned to Thoreau because they could not be sold. *Walden* initially did little better. Thoreau also saw about a dozen of essays and several poems published before he died of tuberculosis in 1862, a couple of months short of forty-five years old. After his death Thoreau was "discovered." Relatives and friends arranged for the posthumous publication of several books, poems, essays, and his journals, all of which he left to posterity in manuscript form. And *Walden* was reprinted, and reprinted, and reprinted . . . and remains in print to this day.

The following short selection is taken from Chapter 2 of *Walden*; it gives a sense of Thoreau's style, interests, and approach to nature.

I went to the woods because I wished to live deliberately, to front only the essential facts of life, and see if I could not learn what it had to teach, and not, when I came to die, discover that I had not lived. . . .

. . . Our life is frittered away by detail. An honest man has hardly need to count more than his ten fingers, or in extreme cases he may add his ten toes, and lump the rest. Simplicity, simplicity, simplicity! I say, let your affairs be as two or three, and not a hundred or a thousand; instead of a million count half a dozen, and keep your accounts on your thumb-nail. In the midst of this chopping sea of civilized life, such are the clouds and storms and quicksands and thousand-and-one items to be allowed for, that a man has to live, if he would not founder and go to the bottom and not make his port at all, by dead reckoning, and he must be a great calculator indeed who succeeds. Simplify, simplify. Instead of three meals a day, if it be necessary eat but one; instead of a hundred dishes, five; and reduce other things in proportion. Our life is like a German Confederacy, made up of petty states, with its boundary forever fluctuating, so that even a German cannot tell you how it is bounded at any moment. The nation itself, with all its so-called internal improvements, which, by the way are all external and superficial, is just such an unwieldy and overgrown establishment, cluttered with furniture and tripped up by its own traps, ruined by luxury and heedless expense, by want of calculation and a worthy aim, as the million households in the land; and the only cure for it, as for them, is in a rigid economy, a stern and more than Spartan simplicity of life and elevation of purpose.

Questions

1. Based on the short quotation above, or on your previous readings of Thoreau's works, why do you think that certain environmentalists find Thoreau's writing so appealing and inspirational? How are Thoreau's sentiments applicable to the notions of "appropriate technology," "decentralization," and "local sustainability" found among certain members of the Green movement?

2. Thoreau was living during the middle of the nineteenth century and strongly criticized civilized life as being too rushed and complicated. How do you think he would view American civilization 150 years later? Would he necessarily condemn it?

3. Thoreau's work has been criticized as consisting mostly of style, but containing little real substance. Some people regard Thoreau as a "cult figure" who should have remained relatively obscure, as indeed he was during his lifetime. How do you respond to these comments? Do you agree or disagree with them? Why?

Source

Thoreau, Henry David, 1854, *Walden, or Life in the Woods*. Boston: Ticknor and Fields. [Reprinted many times; one widely available version is published by Signet Classic, Penguin Books, New York, 1960, 1980, and includes an afterward and bibliography by Perry Miller.]

-55-

The Life and Death of Chico Mendes

Chico Mendes (Francisco Alves Mendes Filho, born 1944) was murdered outside the back door of his house in Acre, Xapuri, Brazil, on December 22, 1988. At the time of his death he was not only president of the Xapuri Rural Workers Union, but also a living symbol of the struggle to save the Amazonian rainforest. Upon his death he was hailed as a martyr to the environmental cause.

Mendes was first and foremost a seringueiro, or rubber tapper, as his father and grandfather before him had been. Rubber tappers live in remote areas of the tropical rainforest and tap the wild rubber trees for latex from which natural rubber can be made. Each tapper may "own" several hundred trees, visiting each about once a week. V-shaped cuts are made into the bark of the trees and in this manner the latex sap is collected. If done properly the tapping does not hurt the trees; some rubber trees have been continuously tapped for fifty years or more. Rubber tappers also generally grow a few crops around their house, and collect wild nuts, fruits, and other products from the tropical forest. All in all, the rubber tappers have learned to live sustainably with the tropical rainforests, extracting no more resources than can be replaced.

Beginning in the 1960s the Brazilian government began a process of heavily developing and logging the forests. Wood from the forests could be sold, helping to reduce the country's enormous debt. In the 1970s and 1980s major highway projects were undertaken in the Amazon region, opening up the forest to loggers, ranchers, land speculators, and settlers. Ranchers would clear and burn the forests, turning the land into pasture to raise cattle to supply cheap beef to foreign markets (including fast food chains in America). The burgeoning population of Brazil, most of whom lived in crowded coastal cities, could be spread throughout the sparsely populated Amazonia regions. But logging, ranching, and growing crops in a "modern" fashion are nonsustainable activities in the rainforest. The soil is thin and nutrient-poor. Once the natural vegetation cover is removed what little topsoil exists is easily destroyed and washed away by the heavy rains. In slash-and-burn farming techniques, utilized by many of the new settlers, the ashes of the burnt vegetation fertilizes the soil for one growing season; afterwards it quickly becomes virtually sterile. After a few years the totally destroyed land must be abandoned; ranchers and settlers move on. Land speculators can accumulate huge expanses of used land for little if any cost.

The destruction of the rainforest came with a major human cost. Both the indigenous peoples (the native "Indians") and the rubber tappers were displaced from their homes. As early as 1966 Chico Mendes was standing up for the rights of the tappers, inspired in part by a Marxist tapper named Euclides Fernandez Tavora who had taught Mendes to read and write. In the 1970s and 1980s Mendes became a major leader of the rubber tappers. In 1975 a union of rubber tappers was organized, Mendes being elected the secretary-general. By the late 1970s the rubber tappers had developed some real power and political strategy. Their basic goal was to protect their way of life, but to do so they had to protect the rainforest upon which they depended.

The rubber tappers developed a basic plan of nonviolent confrontation, the empate, which tended to yield positive results. Whenever the rubber tappers learned of an area in the forest that was to be cleared they would gather there and directly confront the loggers (poorly paid workers hired by the logging companies or ranchers to clear the forest). The rubber tappers discussed the situation with the loggers, making it clear that the loggers would destroy their own future by destroying the forest. When all went according to plan, as it often did, the loggers would be convinced to give up their equipment and go home. The rubber tappers would then destroy the logging camp, effectively stopping (at least temporarily) the destruction of the forest in the immediate area. Naturally, such empates were a real problem for ranchers and developers. As they continued and the rubber tappers gained ever more power, the ranchers actively fought back. Violence escalated and hundreds of people died in land disputes between rubber tappers and ranchers; ultimately the violence would take Mendes.

In 1981 Mendes became president of the Xapuri Rural Workers Union, in 1985 he helped plan a national meeting of rubber tappers, and in 1986 he ran unsuccessfully for a seat in the Brazilian legislature as a candidate espousing forest protection (Mendes came very close to winning, and there were allegations that voting fraud ultimately defeated him). In the 1980s Mendes was "discovered" by English and American environmentalists who promoted him as a symbol of the grass-roots struggle to save the Amazonian rainforest. Mendes and his fellow rubber tappers had come up with the idea of "extract reserves," areas of the rainforest that would be preserved for traditional, sustainable uses as practiced by the rubber tappers and native indigenous peoples. Mendes's American supporters latched on to this idea and promoted Mendes in the international media as a folk hero and a rallying point for the international movement to save the rainforests. In 1987 Mendes's American backers arranged for him to visit the United States where he spoke with members of Congress and convinced them not to appropriate $200 million to the Inter-American Development Bank for road building, leading to more deforestation, in Brazil. Mendes received the United Nations Environment Program's Global 500 award, and back in Brazil he was able to meet with the Interior Ministry. In 1988 Brazil agreed to establish several extractive reserves.

Mendes's activities engendered more and more animosity from the ranchers, especially the local ranchers where he lived and worked. The Alves da Silva family, a large ranching family, was particularly affected by Mendes's activities. An area they claimed became an extractive reserve, and their plans to expand their holdings were foiled. Violence escalated, and by mid-June, 1988, several associates closed to Mendes had been killed in Xapuri. Death pronouncements were made against Mendes directly. Mendes carried a gun for protection and had police officers assigned to guard him, although he still espoused nonviolence. The end came on December 22, 1988, when Mendes was shot in the chest with a shotgun as he went out his back door. Two years later two members (father and son) and a ranchhand of the Alves da Silva family were convicted of the murder, although the father's conviction was overturned in March 1992. However, other charges of murder were pending and he remained in the poorly secured prison until father and son escaped in February 1993. Beyond those directly implicated in Mendes's murder, there have been continuing allegations that many major ranchers and land speculators may have encouraged, or even paid, the da Silvas to kill Mendes.

Questions

1. What conditions led to the wholesale destruction of the Amazonian rainforest? How was this destruction promoted by foreign (outside of Brazil) governments, markets, and financial institutions?

2. Mendes openly stated that his main concern was the protection of the rubber tappers' way of life, not the protection of rainforest per se. Why did American environmentalists and conservationists take such an active interest in promoting Mendes and his struggle on behalf of the rubber tappers? Do you think they were using Mendes for their own purposes?

3. It has been suggested that even if members of the da Silva family were directly to blame for Mendes's death, indirectly the desires of consumers in the developed countries, such as Canada and the United States, are partly responsible for his death. It is such consumers, some contend, who demand fancy tropical woods for expensive furniture and cheap beef that can be sold in fast-food outlets. Do you think there is any truth to such an argument? Are such critics of the developed countries simply seeking to make consumers feel guilty? What "hidden agenda" might such critics have? Do they perhaps want to increase the sales of local American or Canadian products?

Sources

Cunningham, William P., and Barbara Woodworth Saigo, 1992, *Environmental Science: A Global Concern.* Dubuque, Iowa: Wm. C. Brown Publishers.

Newton, Lisa H., and Catherine K. Dillingham, 1994, *Watersheds: Classic Cases in Environmental Ethics.* Belmont, California: Wadsworth Publishing Company.

Revkin, Andrew, 1990, *The Burning Season.* Boston: Houghton Mifflin.

Shoumatoff, Alex, 1990, *The World is Burning.* Boston: Little, Brown.

-56-

Catastrophic Environmental Predictions from the First Earth Day (1970)

In the early days of the modern environmental movement many activists made predictions as to what would happen to Earth and human society if nothing was done to stem the growing environmental crises. One of the most famous prognosticators was (and still is) Dr. Paul R. Ehrlich, an ecologist and professor of biology at Stanford University. In 1968 Ehrlich's widely-read book warning of the problems of global overpopulation, *The Population Bomb*, was first published; twenty-two years later Ehrlich, with his wife Anne H. Ehrlich, published the follow-up book, *The Population Explosion*. In many other books and articles Ehrlich has speculated as to the future of the planet if we humans do not change our environmentally destructive ways.

In the September 1969 issue of the magazine *Ramparts* Ehrlich presented one hypothetical scenario, a very bleak picture of the Earth in 1979-1980. Extrapolating air pollution trends of the 1960s into the 1970s, he predicted incident solar radiation of the Earth's surface would be so reduced by air pollution that all vegetation would detrimentally affected. Furthermore, hundreds of thousands of American would die each year as a direct effect of air pollution. Due to both air pollution and chemical contaminants in the water and soil, by the late 1970s the life expectancy of the average American born after 1946 (when DDT first came into use) was set at 49 years.

In Ehrlich's scenario, the whaling industry died in 1973, and the Peruvian anchovy industry, as well as many other fisheries, had collapsed by 1977. Indeed, Ehrlich predicted that the annual ocean fish yield would be reduced to 30 million metric tons in 1977. By the late 1970s the entire planet would be heavily polluted with chlorinated hydrocarbons (such as DDT), and by the late summer of 1979 the ocean would be officially declared "dead."

On land there would be extreme water rationing, and hepatitis and dysentery rates would skyrocket, not only in developing countries but also in developed countries like the United States and Canada. By the early 1970s the Green Revolution would have collapsed, meaning a loss of agricultural output. Pests (such as insect pests and rodents) would proliferate even as hunger swept the human population. Food prices would become astronomical and tens of millions of people would starve in famines of unheard-of proportions. Adding to the problems, a shift in the jet stream would turn the Midwestern breadbasket region into a desert.

Ehrlich predicted that in the United States by about 1974 a national consensus would develop that the only way to solve the environmental problems of the world would be population control--among the developing nations, that is. This idea did not go over well among the governments of the developing countries, and the world's population continued to grow.

Fortunately, most of Ehrlich's predictions for the 1970s were very far off the mark. DDT and most other chlorinated hydrocarbons have been banned in the United States. Clean air, water, and waste legislation of the 1970s did much to alleviate pollution, as did the establishment of the Environmental Protection Agency in 1970. The air was not overwhelmed by pollution, at least in the way that Ehrlich envisioned, and the oceans did not die. Ocean fish yields of the 1970s were nearly double Ehrlich's predictions.

On land, the Green Revolution did not collapse in the 1970s; world grain production, grain yields, and grain per person continued to increase throughout the 1970s and most of the 1980s. Water pollution, and the implementation of water rationing, have occurred--but not to the extent that Ehrlich envisioned. American life expectancies did not plummet. In the United States life expectancy at birth in 1970-1975 was 71.3 years and by 1990-1995 it had risen to 75.9 years (in Canada the life expectancies were even higher, 73.1 and 77.4 respectively).

As for Ehrlich's political predictions, the U.S. never really developed the national consensus concerning population control that he predicted. Granted, many citizens are concerned about population control, but in the 1980s the Reagan administration effectively put a damper on any American leadership concerning global population control.

Many things happened in the 1970s and 1980s that Ehrlich missed completely. Global warming caused by a type of air pollution--primarily carbon dioxide--and the destruction of the ozone layer--caused by the release of CFCs--both became major environmental issues of the 1970s and 1980s. Nor did Ehrlich anticipate the rise and decline of the nuclear power industry or the interest in alternative energy sources and energy conservation during the 1970s and beyond (although ground was lost in this area during the 1980s).

Questions

1. Given that he was writing for a popular audience in a science fiction vein, how seriously do you think Ehrlich's 1969 predictions were meant to be taken?

2. Environmentalists are sometimes criticized for being full of "doom and gloom," making dour predictions that rarely come true. Do you think this is a just accusation? Do you think such incorrect predictions help or hurt the environmental cause?

3. When is it useful to attempt to predict the future? (Consider benefit-cost analyses that must include predictions of future outcomes given certain actions.) Under what circumstances should serious predictions be made?

Sources

Ehrlich, Paul R., 1968, *The Population Bomb*. New York: Ballantine.

Ehrlich, Paul R., 1969, "Eco-Catastrophe!" *Ramparts* (September 1969). Reprinted in Garrett De Bell, editor, 1970, *The Environmental Handbook: Prepared for the First National Environmental Teach-In*. New York: Ballantine Books, pp. 161-176.

Ehrlich, Paul R., and Anne H. Ehrlich, 1990, *The Population Explosion*. New York: Simon and Schuster.

-57-

Tree Spiking

Tree spiking is a technique used by some members of the radical fringes of the environmental movement, such as some (but not all) Earth First!ers. Spiking consists of driving nails or spikes into trees (which does not hurt the trees) so as to stop or discourage lumberers from cutting down the trees. If a lumber company cuts a spiked tree it may damage equipment upon hitting a spike. Sawblades may snap and go flying. A sawblade in a lumber mill may cost up to $3000, so the economic losses from dealing with spiked trees can be substantial. Furthermore, lumberjacks and millworkers could be seriously injured, or even killed, in the process.

Many different types of spikes can be used to sabotage lumbering areas. Most commonly metallic nails or spikes, available from any good hardware store, in the range of four to eleven inches long are driven into trees. In order to make it more difficult to remove them, nails with a spiral threading are sometimes used, and/or the heads of the nails are removed. Steel welding rods can also be used, as well as spikes made of rock or specially baked, hard ceramics. In order to insert these types of spikes a hole is drilled into the tree first. The rock and ceramic spikes are used so that they cannot be found using metal detectors.

Tree spiking is not new: as early as 1875 laws prohibiting tree spiking were passed in California, but it was first well-publicized and popularized in the middle 1980s by members of Earth First! During this time period a number of spikings were carried out in the Pacific Northwest so as to discourage, or at least stall, timber sales. Common targets are old-growth forest or very mature second-growth trees that are in imminent danger of being cut. Even if the spiking does not stop the lumbering altogether, it slows it down and costs the companies and forest service considerable amounts of money, money that tree spikers figure would otherwise go toward the destruction of even more trees. It is claimed by Earth First!ers that timber sales on public lands have often been quietly cancelled after it was learned that the trees had been spiked. An early spiking incident was the placing of 11,000 spikes into spruce and hemlock trees on Meares Island, off the British Columbian coast of Canada, during the winter of 1984-1985. This was done in response to official plans to cut 90% of the trees on the 21,000 acre island over the next twenty years.

According to the ethics adopted by most tree spikers, the spiking is intended only to save the trees from being lumbered. The explicit intention is to avoid the injury of any lumberjacks or millworkers. Therefore, tree spikers have almost invariably announced exactly what stands of trees they have spiked. Very few lumberers or millworkers have ever been injured due to spiking; one California millworker was severely injured, however, when the saw of the machine he was operating hit an eleven-inch spike. In this case, the mill claims it was never warned that the trees might have been spiked. An official investigation attributed this particular spiking to a lone, right-wing suspect rather than to left-wing environmentalists. However, the suspect was never charged and the incident did bring negative publicity to tree spiking environmentalists. Even if this particular injury-causing spiked tree was not due to environmentalists, many commentators argued that Earth First!ers who spike trees must bear some responsibility for the publicizing of such spiking, potentially encouraging others to do the same. Given this context, some Earth First!ers have explicitly renounced tree spiking because it is the workers, not the presidents and chief executive officers, of lumbering companies who will be directly injured if spike warnings go unheeded and spiked trees are lumbered and milled.

Questions

1. Do you believe that tree spiking can be an effective means of discouraging timber sales and lumbering? Even if it is effective, is it justifiable?

2. Even if ecosaboteurs announce that trees have been spiked before lumbering begins, and the timber sales are therefore cancelled, what about the future? Conceivably fifty or a hundred years from now someone might be seriously injured or killed when their saw hits a spike that was buried in a different era and long since forgotten. An analogy might be children in 1996 who are maimed or killed when they stumble upon World War I land mines. Who bears the responsibility for such incidents? Should tree spikers be worried about such matters? Or do the immediate concerns of the present outweigh such future considerations?

3. Tree spikers have often been called terrorists, or more specifically ecoterrorists. However, tree spikers counter that terrorists threaten and kill innocent people. Tree spikers always warn the appropriate officials once trees are spiked in order to protect the lives of both people and trees. Many radical environmentalists claim that the real terrorists are the timber company executives who exploit the workers (many accidents occur in the lumber industry) and the environment for their own profit. Who do you think the "terrorists" are?

Source

Scarce, Rik, 1990, *Eco-warriors: Understanding the Radical Environmental Movement.* Chicago: The Noble Press, Inc.

-58-

Legal Rights for Trees and Streams

Various environmentalists have argued that natural objects, such as animal individuals and species, plant species, stands of forest, mountains, streams, entire ecosystems, and the land itself should be accorded legal rights. Such ideas are viewed by some as extremely radical, or perhaps just nonsensical. In the status quo and the popular imagination, with few exceptions, it is primarily human beings who are given legal rights. A tree, for instance, in and of itself does not have any rights although the human owner of the tree might have certain rights--and those rights might be violated if another human illegally damaged or destroyed the tree. At one time not even all human beings were viewed as possessing legal or natural rights; slaves two hundred years ago were denied many rights as a matter of course and "common sense." Today most people would agree that all humans are due certain rights, and the concept of rights is slowly being broadened to include non-humans such as sentient animals. Still, we are a long way from granting legal rights to forests, mountains, and so forth.

Christopher D. Stone, professor of law at the University of Southern California, has cogently argued for the extension of true legal rights to natural objects. Professor Stone does not contend that natural objects should necessarily have the same rights as humans, or that even all aspects of the environment should have the same rights as all other aspects. But he does believe that certain rights should be accorded to certain natural objects, and those natural objects (be they trees or mountains) should be able to seek redress when their rights are violated.

Such a proposition immediately entails practical problems. For one, even if it is agreed what rights the natural objects will possess--perhaps the right to exist without being irreparably damaged by humans--how can such natural objects seek redress if they cannot speak? Professor Stone points out that already many entities are generally and legally acknowledged as having certain rights even if they cannot speak. Such entities include corporations, universities, estates, municipalities, states, nations, human infants, persons deemed incompetent, and so forth (note that all of these are human institutions or human beings which for one reason or another cannot speak for themselves). Who speaks for institutions, estates, infants, the senile, and others who cannot represent themselves? Lawyers, parents, friends, advocates, guardians, etc., do the speaking. In fact, in legal situations many perfectly competent and well-spoken individuals do not speak for themselves--they let lawyers do the talking. Under Stone's proposal, speaking humans could become guardians and spokespersons for non-speaking natural objects.

An example illustrates how this proposal might work. Imagine a mining company in the process of strip mining the side of a mountain. Someone interested in the mountain and its well-being might become concerned that the mining company's activities will result in severe and irreparable damage to the mountain and its ecosystem. The concerned citizen (or perhaps group of citizens, such as environmental organizations like The Sierra Club or the Natural Resources Defense Council, just to name two) might apply to the applicable court to be appointed guardian of the mountain. As guardian of the mountain, the concerned person could then speak for the mountain, seek redress on behalf of the mountain, and be afforded certain privileges, such as access to the mountain in order to monitor the mining operations. An important point of such an approach would be that the guardian would not be acting out of self-interest, but strictly in the interest of the mountain and for the protection of the mountain's rights. If the mining company caused damages for which compensation was awarded by a court, the money would not go to the guardian but rather to the mountain (perhaps to repair and restore the mountain, or into some kind of trust fund to benefit the mountain in the future).

Questions

1. Given current thinking, how realistic do you think it is to accord rights to mountains, trees, and other natural objects? Who would raise objections to such rights? Why? How would such rights conflict with the rights currently accorded to property owners?

2. If we begin to grant legal rights to certain natural objects, does that pave the way for the abolition of private ownership of such natural objects? Compare this to the granting of legal rights to all humans and the abolition of slavery.

3. If legal rights were granted to natural objects, many human guardians would have to be established to speak for and watch after the rights and interests of the objects. How important is it that the "correct" guardian be appointed for a particular natural object? In the hypothetical case of the mountain described above, what if the president of the mining company had been appointed guardian of the mountain? Should the guardian be legally obligated to place the interests of its charge before any self-interest? What might happen when a human guardian dies? Should the courts appoint a new guardian, or might guardianships be passed down from one generation to another? Might guardianships become a new form of "ownership"? Could such problems be avoided by appointing groups or organizations, rather than individuals, as guardians?

Source

Stone, Christopher D., 1972, "Should Trees Have Standing? Toward Legal Rights for Natural Objects." Reprinted in White, James E., 1994, *Contemporary Moral Problems (fourth edition)*. Minneapolis/St. Paul: West Publishing Company, pp. 437-442.

-59-

The Takings Concept: When is Just Compensation to a Property Owner Required?

According to the Fifth Amendment of the United States Constitution, private property may not be taken for public use unless due process is followed and just compensation is paid to the owner. However, exactly what constitutes a "taking" as described in the Constitution is currently a subject of intense debate. Historically, if a government appropriated a parcel of land for public use (perhaps to build a new school or road), this would constitute a clear example of a taking and the owner would be entitled to fair market value for the property. However, if for the good of the public a property owner is restricted in his or her use of private land, does this also constitute a taking? Governments have long imposed regulations for the public good without compensating private owners who may experience some financial loss or burden. An example of such a restrictive regulation might be a ban on dumping hazardous wastes on private property--traditionally this would not constitute a governmental taking even if the property owner could clearly demonstrate that it would impose an economic burden (perhaps by reducing the value of the land which had been purchased, prior to the enactment of the regulation, for the purpose of erecting a dump). However, a recent movement led by certain business and property owners has argued that the concept of "takings" should be expanded to include financial losses caused by restrictive regulation. In 1992 an important takings case came before the United States Supreme Court: Lucas v. South Carolina Coastal Council.

In 1986 David H. Lucas, a contractor and developer, paid $975,000 for two beachfront residential lots on the Isle of Palms, Charleston County, South Carolina. It was Lucas's intention to erect single-family homes on the lots. In 1988 the South Carolina legislature enacted the Beachfront Management Act which had the effect of banning the erection of any permanent structures on the lots that Lucas had purchased. Lucas argued that this effectively constituted a taking since, as a developer, he was deprived of any reasonable economic use of the property and consequently his property had now become "valueless." The case went to a state trial court and the court sided with Lucas; just compensation to be paid to Lucas was set at slightly over $1.2 million. However, the state took the case to the Supreme Court of South Carolina which reversed the decision. Lucas then appealed the case to the United

States Supreme Court which decided to hear the arguments.

The Supreme Court of South Carolina had argued that if a regulation regarding the use of property is intended "to prevent serious public harm" then no compensation is due to the owner. In this case, the serious public harm was seen as erosion of beaches on coastal barrier islands--erosion which is promoted by the building of permanent dwellings. The proposed activities (building homes) of Lucas were regarded as a "nuisance-like activity." It was also argued that even if Lucas was not allowed to build homes on the property, that did not deny him all potential economically beneficial uses of the property. Alternatively, the property might be used as a bathing beach, a campground, a boat dock, a nature preserve, or some other enterprise. Lucas himself came under fire: it was asserted that the Wild Dune Development Company, of which Lucas was a part owner, was notorious for the unstable nature of much of the property it owned. Over the previous forty years, it was argued, Wild Dune property was often flooded; local authorities had to issue twelve emergency sandbagging orders to protect Wild Dune property just between 1981 and 1983.

In Lucas's favor was the fact that the first trial court had found the act to render Lucas's parcels "valueless," at least for his purposes. When Lucas purchased the property it was clearly and legally zoned for single family residential construction without any restrictions imposed by the state of South Carolina, Charleston County, or the town of Isle of Palms. It was argued that the South Carolina Act did not so much prevent Lucas from "harming" the property (by building houses) as it promoted the "benefit" (a benefit in the public interest) of not having houses on the property--a "benefit" that would be gotten at Lucas's expense if he did not receive just compensation for the taking.

The issues of whether homes or empty beaches are more beneficial, and whether or not housing construction is harmful, are matters of judgement heavily dependent on one's individual ideals, goals, and beliefs. What Lucas proposed to do--build houses--was clearly not a public threat or nuisance on the same order as building a dump, for instance. In favor of Lucas, it was stated that "[T]he problem [in this case] is not one of noxiousness or harm-creating activity at all; rather it is a problem of inconsistency between perfectly innocent and independently desirable uses."

Questions

1. Evaluate the argument that the proposed activities of Lucas (building two houses) can be viewed as a nuisance-like activity that would cause public harm. Why weren't such activities viewed this way in 1986 when Lucas purchased the property?

2. If you were Lucas, would you be upset at losing nearly a million dollars when you had purchased property in good faith with the understanding that you would be allowed to build on it? Would you take the case to court?

3. Evaluate the argument that building houses and protecting beachfront property by the restriction of building activities are both "perfectly innocent and independently desirable uses" of the property.

4. If you were a member of the United States Supreme Court, how do you think you would rule on this case? Justify your position.

The Supreme Court's Decision

The majority of the justices of the U.S. Supreme Court ruled in favor of Lucas, although a minority dissented. This ruling has been hailed as a major victory for advocates of an expanded takings concept and as a major defeat for environmentalists. It is important to note that this ruling in no way forbids any legislative regulatory activities, but it does potentially greatly expand the situations in which the government must compensate owners of private property for takings. In effect, this means it could (and almost surely will) become much more costly to enforce many environmental regulations. In the case of South Carolina and the former Lucas property, after the state was forced to purchase the two house lots from Lucas officials rethought the whole situation. They finally decided to allow housing on the very same two lots after all, and sold the lots to developers in order to recover the expenses paid to Lucas. So in this case it was decided that if compensation would be required for such takings then it would be too expensive to enforce the original act; consequently, the South Carolina authorities backed down.

Sources

Allen, Scott, 1995, "Land Grab: Property rights challenges are raising the ante in environmental protection." *The Boston Globe* (December 11, 1995): pp. 29-31.

U.S. Supreme Court Ruling, 1992, "Lucas v. South Carolina Coastal Council." In *Introduction to Environmental Law and Policy* (by C. M. Valente and W. D. Valente, 1995), pp. 299-303. Minneapolis/St. Paul: West Publishing Company.

-60-

Easterbrook's Concept of a New Nature

Gregg Easterbrook, at the end of his book *A Moment on the Earth*, has suggested that in the future we may have a "New Nature," one designed by human intellect. This may seem a radical, even repulsive idea at the moment, but Easterbrook contends that a "New Nature, modified by men and women, is coming. It cannot be stopped, nor should it. The issue that matters is how to make the New Nature good rather than bad" (p. 668 of *A Moment*). Indeed, one of the basic lessons of environmental science today is that humanity is a major force that has already transformed the planet, even if in many cases inadvertently, and will continue to transform the planet. Easterbrook argues that this New Nature should be by design rather than simply allowed to occur--we cannot stop change, but we now have it within our power to direct it.

What does Easterbrook envision for his ideal New Nature? He suggests a number of specific points:

1) Through large scale genetic intervention, the New Nature could end the predation of animals by other animals. This, according to Easterbrook, would set in place an era of cooperation between animals and essentially correct a flaw in nature--namely the concept of a predator. "[P]resent day predators could become herbivores, continuing to live as wild bears or wolves or weasel, except leaving out the gruesome part. Nature might long for such a reform" (p. 671).

2) In the New Nature humans may still eat meat, but they will not kill animals to obtain the meat. Using modern culturing and genetic engineering techniques, beef, pork, chicken, fish, and other animal cells and meat could be grown in laboratories without having to raise and slaughter the actual animals. This would be much more humane, and resources would not be wasted on the bones, intestines, brain, and other parts of animals that are not utilized.

3) Easterbrook suggests that perhaps through genetic manipulation the entire human race could be rendered genetically incapable of desiring to kill one another, thus rendering humanity a much more peaceful species.

4) In the New Nature there might be no such thing as species extinctions. Without explanation, Easterbrook suggests that humans may figure out a way to preserve all species.

5) In the New Nature envisioned by Easterbrook all disease would be eradicated--not just disease among humans, but among animals and plants as well.

6) Humans would build a network around the Earth to detect and protect the planet from extraterrestrial invaders, such as comets, asteroids, large meteorites, and the like. Using missiles, lasers, or other devices such danger would be either destroyed or pushed out of potential collision courses with Earth. The giant asteroid that is suspected by some of causing the extinction of the dinosaurs would never have a chance in the New Nature.

7) In the New Nature of Easterbrook there might be no more aging--no aging for people, and perhaps not for plants and animals either. This might be accomplished by genetic engineering. A world without aging would not be the same as a world without death; organisms could be killed by accident (or by disease or violence, if these still exist).

8) Today most of the Sun's energy is "wasted"--it streams out into deep space. In the New Nature most or all of the Sun's output might be captured and used by humans or nature. Easterbrook resurrects physicist Freeman Dyson's idea that the matter composing one of the lifeless planets of the outer solar system (perhaps Pluto, or Pluto combined with matter accumulated from miscellaneous comets, asteroids, and so forth) could be used to construct a huge reflective sphere around the remaining solar system. Such a Dyson sphere could capture virtually 100% of the Sun's energy, energy that could be used to make such planets as Mars, Neptune, and Uranus habitable for life.

9) Finally, Easterbrook suggests that human consciousness, currently limited (as far as can be demonstrated) to the chemical and electrical patterns found in the human brain, may in the future be transferable to other forms of matter (other thinkers have suggested that a specific human consciousness, what some would call a soul, could be sustained on an advanced computer chip). Through technology, we could ensure that a specific person's consciousness would continue to live even if the body ceased to exist. Easterbrook even suggests that "perhaps someday some form of technology might even sustain patterns of consciousness in a noncorporeal manner" (p. 675). That is, human consciousness could exist without the presence of physical matter.

Easterbrook does not imagine that his New Nature will come quickly--it may take hundreds or even thousands of years to implement--but he does think it is possible. Easterbrook contends that his vision of a New Nature would benefit both humanity and the planet as a whole.

Questions

1. Easterbrook's concept of a New Nature depends heavily on technological advancements, especially genetic engineering. If it were possible, would you approve of such artificial genetic manipulation on a global scale?

2. Easterbrook's fourth point cited above, the end of extinction, is followed by his concept of the eradication of all disease. Are these two points mutually incompatible? How can we

eradicate disease and preserve all species simultaneously? Aren't many "diseases" also "species"? Furthermore, are we really preserving species from extinction if we are changing them radically through genetic intervention (for instance, turning carnivores into herbivores, as Easterbrook suggests)?

3. If people, and perhaps even plants and animals, do not age in the New Nature, should everyone be sterile? Will there be a place for children? Would there be a need to support consciousness outside of the human brain if nobody ages?

4. Do you think you would want to live in a world encompassed by Easterbrook's New Nature? Why or why not? Once the New Nature had been in existence for a few centuries, do you think it would come to be accepted as the "way things should be"? If you could go back in time and describe late twentieth century technological society to a typical person of the early 17th century, do you think he or she would believe you? Do you think he or she would want to join us in the present day, or might our New Nature seem "wrong" to a seventeenth-century mind?

Source

Easterbrook, Gregg, 1995, *A Moment on the Earth: The Coming Age of Environmental Optimism*. New York: Viking.